Signal Processing for Image Enhancement and Multimedia Processing

T0189452

MULTIMEDIA SYSTEMS AND APPLICATIONS SERIES

Consulting Editor

Borko Furht
Florida Atlantic University

Recently Published Titles:

Visit the series on our website: www.springer.com

Signal Processing for Image Enhancement and Multimedia Processing

edited by

Ernesto Damiani
University of Milan
Italy

Albert Dipanda
University of Bourgogne
France

Kokou Yétongnon
University of Bourgogne
France

Louis Legrand
University of Bourgogne
France

Peter Schelkens
Vrije University of Brussels
Belgium

Richard Chbeir
University of Bourgogne
France

 Springer

Ernesto Damiani
Università Milano-Bicocca
Dipto. Tecnologie dell'Informazione
via Festa del Perdono,7
20122 MILANO, ITALY
damiani@dti.unimi.it

Albert Dipanda
Université de Bourgogne
LE2I-CNRS
Aile de l'ingénieur
21000 Dijon, FRANCE
adipanda@u-bourgogne.fr

Kokou Yétongnon
Université de Bourgogne
LE2I-CNRS
Aile de l'ingénieur
21000 Dijon, FRANCE
kokou@u-bourgogne.fr

Louis Legrand
Université de Bourgogne
LE2I-CNRS
21000 Dijon, FRANCE
Louis.legrand@u-bourgogne.fr

Peter Schelkens
Vrije Universiteit Brussel
Dept. Electronics and Info. Processing (ETRO)
Pleinlaan 2
1050 BRUXELLES, BELGIUM
Peter.Schelkens@vub.ac.be

Richard Chbeir
Université de Bourgogne
LE2I-CNRS
Aile de l'ingénieur
21000 Dijon, FRANCE
Richard.chbeir@u-bourgogne.fr

Signal Processing for Image Enhancement and Multimedia Processing
edited by Ernesto Damiani, Albert Dipanda, Kokou Yétongnon,
Louis Legrand, Peter Schelkens and Richard Chbeir

ISBN-13: 978-1-4419-4442-9   eISBN-13: 978-0-387-72500-0

Printed on acid-free paper.
© 2008 Springer Science+Business Media, LLC
Softcover reprint of the hardcover 1st edition 2008

9 8 7 6 5 4 3 2 1

springer.com

Preface

Traditionally, signal processing techniques lay at the foundation of multimedia data processing and analysis. In the past few years, a new wave of advanced signal-processing techniques has delivered exciting results, increasing systems capabilities of efficiently exchanging image data and extracting useful knowledge from them. Signal Processing for Image Enhancement and Multimedia Processing is an edited volume, written by well-recognized international researchers with extended chapter style versions of the best papers presented at the SITIS 2006 International Conference.

This book presents the state-of-the-art and recent research results on the application of advanced signal processing techniques for improving the value of image and video data. It also discusses feature-based techniques for deep, feature-oriented analysis of images and new results on video coding on time-honored topic of securing image information. Signal Processing for Image Enhancement and Multimedia Processing is designed for a professional audience composed of practitioners and researchers in industry. This volume is also suitable as a reference or secondary text for advanced-level students in computer science and engineering.

The chapters included in this book are a selection of papers presented at the Signal and Image Technologies track of the international SITIS 2006 conference. The authors were asked to revise and extend their contributions to take into account the many challenges and remarks discussed at the conference. A large number of high quality papers were submitted to SITIS 2006, demonstrating the growing interest of the research community for image and multimedia processing.

We acknowledge the hard work and dedication of many people. We thank the authors who have contributed their work. We appreciate the diligent work of the SITIS committee members. We are grateful for the help, support and patience of the Springer publishing team. Finally, thanks to Iwayan Wikacsana for his invaluable help

Dijon, Milan
July 2007

Ernesto Damiani
Kokou Yetongnon
Albert Dipanda
Richard Chbeir

Contents

Part II Texture Analysis and Feature Extraction

Part III Face Recognition and Shape Analysis

List of Contributors

Abu Sayeed Md. Sohail
Concordia Institute for Information
Systems Engineering (CIISE)
Concordia University, 1515 St.
Catherine West,
Montreal, Canada
a_sohai@encs.concordia.ca

Albert Dipanda
LE2i -, Bourgogne University,
Dijon, France
albert.dipanda@u-bourgogne.fr

Aleksandra Pižurica
Ghent University,
St-Pietersnieuwstraat 41, B-9000,
Ghent, Belgium
aleksandra.pizurica@telin
.ugent.be

Amine Bermak
Department of Electrical and
Electronic Engineering,
Hong Kong University of Science
and Technology Clear Water Bay,
Kowloon, Hong Kong, SAR
bermak@ieee.org

Ammar Bouallègue
SYSCOM Lab, ENIT,
Tunis,Tunisia
ammar.bouallegue@enit.rnu.tn

Angel D. Sappa
Computer Vision Center
Edifici O Campus UAB
08193 Bellaterra,
Barcelona, Spain
angel.sappa@cvc.uab.es

Atul Negi
Dept. of CIS, University of Hyderabad,
Hyderabad, India
atulcs@uohyd.ernet.in

Benoit Rat
EPFL Ecole Polytechnique Federale
de Lausanne,
Lausanne, Swiss
benoit.rat@epfl.ch

Boris X. Vintimilla
Vision and Robotics Center
Dept. of Electrical and Computer
Science Engineering, Escuela
Superior Politecnica del Litoral
Campus Gustavo Galindo, Prospe-
rina,
Guayaquil, Ecuador
boris.vintimilla@espol.edu.ec

Chen Shoushun
Department of Electrical and
Electronic Engineering,
Hong Kong University of Science
and Technology, Clear Water Bay,
Kowloon, Hong Kong, SAR
dazui@ust.hk

D. Sudhir Kumar
IDRBT, Castle Hills, Road No 1,
Masab Tank,
Hyderabad, India
sudheerdosapati@yahoo.com

Dima Pröfrock
University of Rostock,
Institute of Communications Engi-
neering,
Rostock, Germany
dima.proefrock@uni-rostock.de

Emmanuel Tonye
LETS -, National Higher School
Polytechnic,
Yaounde, Cameroon
tonyee@hotmail.com

Enguerran Grandchamp
GRIMAAG UAG, Campus de
Fouillole
97157 Pointe--Pitre
Guadeloupe, France
egrandch@univ-ag.fr

Erika Müller
University of Rostock,
Institute of Communications Engi-
neering,
Rostock, Germany
erika.mueller@uni-rostock.de

Ernesto Damiani
Department of Information Technology, University of Milan
via Bramante, 65 - 26013,
Crema (CR), Italy
damiani@dti.unimi.it

Fabrice Mairesse
Université de Bourgogne, Le2i UMR
CNRS 5158, Route des plaines de
l'Yonne, BP 16,
Auxerre, France
Fabrice.Mairesse@u-bourgogne.fr

Fabrice Meriaudeau
Université de Bourgogne - Le2i, 12
rue de la Fonderie,
Le Creusot, France
Fabrice@
iutlecreusot.u-bourgogne.fr

Faouzi Soltani
Laboratoire Signaux et Systèmes de
Communication,
Université de Constantine,
Constantine, Algeria.
f.soltani@caramail.com

Farid Boussaid
School of Electrical Electronic and
Computer Engineering,
The University of Western Australia,
Perth, Australia
boussaid@ee.uwa.edu.au

Gerald J. F. Banon
National Institute for Space Research
(INPE) Av. dos Astronautas,
São José dos Campos, Brazil
banon@dpi.inpe.br

Guaraci J. Erthal
National Institute for Space Research
(INPE) Av. dos Astronautas,
São José dos Campos, Brazil
gaia@dpi.inpe.br

Gwanggil Jeon
Department of Electronics and
Computer Engineering, Hanyang
University, 17 Haengdang-dong,
Seongdong-gu,
Seoul, Korea
windcap315@ece.hanyang.ac.kr

Haifa Belhadj
SYSCOM Lab, ENIT,
Tunis, Tunisia
haifa.belhadj@yahoo.fr

Han Li
Department of Computer Science
and Technology,
University of Technology,
Dalian, China
lihan@student.dlut.edu.cn

He Guo
Department of Computer Science
and Technology,
University of Technology,
Dalian, China
guohe@dlut.edu.cn

Irène Foucherot
LE2I, UMR CNRS 5158 Universit
de Bourgogne, Aile des Sciences de
l'Ingénieur, BP 47870
Dijon, FRANCE
iFoucherot@u-bourgogne.fr

Isameddine Boukhriss
Lyon2 University, LIRIS Laboratory,
5 av. Pierre Mendes-France

Lyon, France
isameddine.boukhriss@
univ-lyon2.fr

Jean Vaillancourt
Département d'informatique et
d'ingénierie, Université du Québec
en Outaouais,
C.P. 1250, Succ. B, Gatineau (Qc),
Québec, Canada
jean.vaillancourt@uqo.ca

Jechang Jeong
Department of Electronics and
Computer Engineering, Hanyang
University, 17 Haengdang-dong,
Seongdong-gu,
Seoul, Korea
jjeong@ece.hanyang.ac.kr

Jimin Wang
College of Computer Science and
Technology, Zhejiang University,
Hangzhou, China

Joost Rombaut
Ghent University – TELIN – IPI –
IBBT,
St-Pietersnieuwstraat 41,
Ghent, Belgium
jorombau@telin.ugent.be

Kamel Bensebaa
National Institute for Space Research
(INPE) Av. dos Astronautas, 1758,
São José dos Campos, Brazil
camel@dpi.inpe.br

Kang Sun
College of Computer Science and
Technology, Zhejiang University,
Hangzhou, China
swankong@126.com

Kris Van Schevensteen
University College of Antwerp
Paardenmarkt 92,
Antwerp, Belgium
kris.vanschevensteen@skynet.be

Laure Tougne
Lyon2 University, LIRIS Laboratory
5 av. Pierre Mendes-France
Lyon, France
laure.tougne@univ-lyon2.fr

Leila M. G. Fonseca
National Institute for Space Research
(INPE) Av. dos Astronautas, 1758,
São José dos Campos, Brazil
leila@dpi.inpe.br

Licheng Jiao
Institute of Intelligence Information
Processing and
National Key Lab for Radar Signal
Processing,
Xidian University
Xian, China.
lchjiao@mail.xidian.edu.cn

Lingdi Ping
College of Computer Science and
Technology, Zhejiang University,
Hangzhou, China

Luciano da F. Costa
Universidade de Sao Paulo- IFSC,
Caixa Postal 369
Sao Carlos - SP - Brasil

Ludovic Journaux
LE2I, Universit de Bourgogne, BP
47870 21078
Dijon, France
ljourn@u-bourgogne.fr

Luigi Arnone
STMicroelectronics Advanced
System Research group
Agrate Brianza (MI),
Milan, Italy
luigi.arnone@st.com

M. Sarifuddin
Département d'informatique et
d'ingénierie, Université du Québec
en Outaouais, C.P. 1250, Succ. B,
Gatineau (Qc),
Québec Canada, J8X 3X7
m.sarifuddin@uqo.ca

Marco Anisetti
Department of Information Technology, University of Milan
via Bramante, 65 - 26013, Crema
(CR),
Milan, Italy
anisetti@dti.unimi.it

Mathias Schlauweg
University of Rostock,
Institute of Communications Engineering,
Rostock, Germany
mathias.schlauweg@
uni-rostock.de

Michel Paindavoine
Laboratoire LE2I - UMR-CNRS,
Université de Bourgogne, Aile des
Sciences de l'Ingénieur,
Dijon, France
paindav@u-bourgogne.fr

Mohamed Abadi
GRIMAAG UAG, Campus de
Fouillole
97157 Pointe--Pitre
Guadeloupe, France
mabadi@univ-ag.fr

Mohamad Susli
School of Electrical Electronic and
Computer Engineering,
The University of Western Australia,
Perth, Australia
abionnnn@gmail.com

Mourad Barkat
Department of Electrical Engineering, American University of Sharjah,
Sharjah, United Arab Emirates.
mbarkat@aus.edu

Munaga. V. N. K. Prasad
IDRBT, Castle Hills, Road No 1,
Masab Tank,
Hyderabad, India
mvnkprasad@idrbt.ac.in

Narcisse Talla Tankam
LE2i -, Bourgogne University,
Dijon, France
narcisse.talla@u-bourgogne.fr

Olivier Laligant
Universite de Bourgogne - Le2i, 12
rue de la Fonderie
Le Creusot, France
o.laligant@
iutlecreusot.u-bourgogne.fr

P. Manoj
IDRBT, Castle Hills, Road No 1,
Masab Tank,
Hyderabad, India
pmanoj@mtech.idrbt.ac.in

Peter Schelkens
Vrije Universiteit Brussel
Brussel, Belgium
peter.schelkens@vub.ac.be

Pierre Gouton
LE2I, UMR CNRS 5158 Universit
de Bourgogne, Aile des Sciences de
l'Ingénieur,
Dijon, France
pgouton@u-bourgogne.fr

Prabir Bhattacharya
Concordia Institute for Information
Systems Engineering (CIISE)
Concordia University,
Montreal, Canada
prabir@ciise.concordia.ca

Rokia Missaoui
Département d'informatique et
d'ingénierie, Université du Québec
en Outaouais,
C.P. 1250, Succ. B, Gatineau (Qc),
Québec, Canada
rokia.missaoui@uqo.ca

Ruchan Dong
Institute of Intelligence Information
Processing and
National Key Lab for Radar Signal
Processing, Xidian University
Xian, China.
ruchandong@hotmail.com

Rym Haj Ali
Ecole Superieure des Communications de Tunis
rym.elhadjali@gmail.com

Sebastiaan Van Leuven
University College of Antwerp
Paardenmarkt 92, B-2000, Antwerp,
Belgium
sebastiaan.vanleuven@gmail.com

Serge Miguet
Lyon2 University, LIRIS Laboratory
Batiment C, 5 av. Pierre Mendes-
France,
Lyon, France
serge.miguet@univ-lyon2.fr

Shuang Wang
Institute of Intelligence Information
Processing and
National Key Lab for Radar Signal
Processing, Xidian University,
Xian, China.

Sofia Ben Jebara
Ecole Superieure des Communications de Tunis,
Tunis, Tunisia
sofia.benjebara@supcom.rnu.tn

Sonia Zaibi
SYSCOM Lab, ENIT,
Tunis, Tunisia
sonia.zaibi@enit.rnu.tn

Stéphane Binczak
Université de Bourgogne, Le2i UMR
CNRS 5158, Aile de l'ingénieur, BP
47870,
Dijon, France
stbinc@u-bourgogne.fr

Tadeusz Sliwa
Université de Bourgogne, Le2i UMR
CNRS 5158, Route des plaines de
l'Yonne, BP 16,
Auxerre, France
Tadeusz.Sliwa@u-bourgogne.fr

Thomas Chalumeau
Universite de Bourgogne - Le2i, 12
rue de la Fonderie,
Le Creusot, France
t.chalumeau@
iutlecreusot.u-bourgogne.fr

Tianyang Liu
Department of Computer Science
and Technology, Dalian
University of Technology,
Dalian, China
liutyang@163.com

Tim Dams
University College of Antwerp
Paardenmarkt 92, B-2000,
Antwerp, Belgium
t.dams@ha.be

Toufik Laroussi
Laboratoire Signaux et Systèmes
de Communication, Université de
Constantine,
Constantine, Algeria
laroussi@yahoo.fr

Valerio Bellandi
Department of Information Technology, University of Milan
via Bramante, 65 - 26013, Crema
(CR),
Milan, Italy
bellandi@dti.unimi.it

Wilfried Philips
Ghent University – TELIN – IPI –
IBBT,
St-Pietersnieuwstraat 41,
Ghent, Belgium
wilfried.philips@telin.ugent.be

Xinyuan Fu
Department of Computer Science
and Technology, Dalian
University of Technology,
Dalian, China
guohe@dlut.edu.cn

Xuezeng Pan
College of Computer Science and
Technology, Zhejiang University,
Hangzhou, China

Yuxin Wang
Department of Computer Science
and Technology, Dalian
University of Technology,
Dalian, China
wyx@dlut.edu.cn

Yvon Voisin
Université de Bourgogne, Le2i UMR
CNRS 5158, Route des plaines de
l'Yonne, BP 16,
Auxerre, France
Yvon.Voisin@u-bourgogne.fr

Zoubeida Messali
Laboratoire Signaux et Systèmes
de Communication, Université de
Constantine,
Constantine, Algeria.
messalizoubeida@yahoo.fr

Part I

Image Restauration, Filtering and Compression

On PDE-based spectrogram image restoration. Application to wolf chorus noise reduction and comparison with other algorithms

Benjamín Dugnol, Carlos Fernández, Gonzalo Galiano, and Julián Velasco

Dpt. of Mathematics, University of Oviedo
c/ Calvo Sotelo s/n, 33007 Oviedo, Spain
dugnol@uniovi.es, carlos@uniovi.es, galiano@uniovi.es, julian@uniovi.es

Summary. We investigate the use of image processing techniques based on partial differential equations applied to the image produced by time-frequency representations of one-dimensional signals, such as the spectrogram. Specifically, we use the PDE model introduced by Álvarez, Lions and Morel for noise smoothing and edge enhancement, which we show to be stable under signal and window perturbations in the spectrogram image. We demonstrate by numerical examples that the corresponding numerical algorithm applied to the spectrogram of a noisy signal reduces the noise and produce an enhancement of the instantaneous frequency lines, allowing to track these lines more accurately than with the original spectrogram. We apply this technique both to synthetic signals and to wolves chorus field recorded signals, which was the original motivation of this work. Finally, we compare our results with some classical signal denoising algorithms and with wavelet based image denoising methods and give some objective measures of the performance of our method. We emphasize that the 1D signal restoration is not the purpose of our work but the spectrogram noise reduction for later instantaneous frequency estimation.

Key words: Spectrogram, time-frequency distribution, noise, partial differential equation, instantaneous frequency, image processing, population counting.

1.1 Introduction

Wolf is a protected specie in many countries around the world. Due to their predator character their protection must be financed from public budgets for farmer's reimbursement of losses and henceforth it is important for authorities to know in advance an estimation of their populations [18]. However, for mammals, few and not very precise techniques are used, mainly based on the recuperation of field traces, such as steps, excrements and so on. In this contribution, we propose what it seems to be a new technique to estimate the population of species which emit some characteristic sounds (howls and barks, for wolves) which consists on identifying how many different voices are emitting in a given recording, task that can be seen as a simplified version of speech recognition, and that we shall approach by instantaneous frequency estimation using time-frequency analysis [9]. The literature on this topic is vast. We refer the reader to, for instance, [4, 6, 12, 13, 15, 16, 19, 20].

However, due to the recording conditions in wilderness, reducing the background unstructured noise in the recorded signal is a necessary step which must be accomplished before any further analysis, being this the main issue of this article. Considering the spectrogram, or any other time-frequency representation, of a signal as an image, we use a PDE image processing technique for edge (instantaneous frequency lines) enhancement and noise reduction based on a regularization of the mean curvature motion equation, as introduced in [1]. See also [7, 8] for related works. There exist a variety of PDE-based models for smoothing and enhancing images that could be used instead, see, for instance, [2, 17]. Other approaches to image denoising, like wavelet analysis, may lead to similar results. Although we do not provide a theoretical analysis, we include some numerical demonstrations using this and other techniques.

1.2 The mathematical model

Let $x \in L^2(\mathbb{R})$ denote an audio signal and consider the Gabor's transform $\mathcal{G} : \mathbb{R}^2 \to \mathbb{C}$ given by

$$\mathcal{G}x(t,\omega) = \int_{\mathbb{R}} x(s)\varphi(s-t)e^{-i\omega s}ds, \qquad (1.1)$$

corresponding to the real, symmetric and normalized *window* φ, with t denoting time and ω frequency. The energy density function or *spectrogram* of x corresponding to the window φ is given by $u_0(t,\omega) = |\mathcal{G}x(t,\omega)|^2$. The regularity of u_0 is that inherited from the window φ, which we assume to be Lipschitz continuous. In particular, u_0 is a bounded function and we may think of it as an *image* and consider its transformation given as the solution $u(\tau,t,\omega)$ of the following problem (*Problem P*), introduced in [1] as an edge-detection image-smoothing algorithm:

$$\frac{\partial u}{\partial \tau} - g(|G_s * \nabla u|)A(u) = 0 \quad \text{in } \mathbb{R}_+ \times \Omega, \qquad (1.2)$$

with the usual no-flow boundary condition $\frac{\partial u}{\partial n} = 0$ on $\mathbb{R}_+ \times \partial\Omega$, and with a given initial image (the spectrogram of x), $u(0,t,\omega) = u_0(t,\omega)$. In (1.2), the diffusion operator is defined as

$$A(u) = (1 - h(|\nabla u|))\Delta u + h(|\nabla u|) \sum_{j=1,\dots,n} f_j\left(\frac{\nabla u}{|\nabla u|}\right)\frac{\partial^2 u}{\partial x_j^2},$$

and the time-frequency domain $\Omega \subset \mathbb{R}^2$ is an open set that we assume to be bounded. Let us remind the properties and meaning of terms in equation (1.2):

Function G_s is a Gaussian of variance s. The variance is a *scale parameter* which fixes the minimal size of the details to be kept in the processed image.

Function g is non-increasing with $g(0) = 1$ and $g(\infty) = 0$. It is a *contrast function*, which allows to decide whether a detail is sharp enough to be kept.

The composition of G_s and g on ∇u rules the speed of diffusion in the evolution of the image, controlling the *enhancement* of the edges and the noise smoothing.

Isotropic and anisotropic diffusion are combined in the diffusion operator, A, smoothing the image by local averaging or enforcing diffusion only on the orthogonal direction to ∇u, respectively. These actions are regulated by $h(s)$, which is nondecreasing with $h(s) = 0$ if $s \leq \epsilon$, $h(s) = 1$ if $s \geq 2\epsilon$, being ϵ the *enhancement* parameter.

1.2.1 Mathematical properties

The following theorem is proven in [1].

Theorem 1. *Let $u_0 \in W^{1,\infty}(\Omega)$. (i) Then, for any $T > 0$, there exists a unique solution, $u \in C([0,\infty) \times \Omega) \cap L^\infty(0,T; W^{1,\infty}(\Omega))$, of Problem P. Moreover,*

$$\inf_{\Omega} u_0 \leq u \leq \sup_{\Omega} u_0 \quad \text{in } \mathbb{R}_+ \times \Omega.$$

(ii) Let v be a solution of Problem P corresponding to the initial data $v_0 \in L^\infty(\Omega)$. Then, for all $T \geq 0$, there exists a constant K which depends only on $\|u_0\|_{W^{1,\infty}}$ and $\|v_0\|_{L^\infty}$ such that

$$\sup_{0 \leq \tau \leq T} \|u(\tau, \cdot, \cdot) - v(\tau, \cdot, \cdot)\|_{L^\infty(\Omega)} \leq K\|u_0 - v_0\|_{L^\infty(\Omega)}. \qquad (1.3)$$

Remark 1. The solution ensured by this theorem is not, in general, a classical solution. The notion of solution employed in [1] is that of *viscosity solution*, which coincides with the classical solution if it is regular enough. Since we will not enter in further discussions about regularity, we refer the reader to [1, 5] for technical details about this notion of solution.

Part *(ii)* of Theorem 1 is specially useful to us for the following reason. Spectrograms of a signal are computed relative to windows, i.e, for each window a different spectrogram (image) is got. Then, the time-frequency characteristics of the signal, like instantaneous frequency, look in a slight different way if two different windows are employed. It, therefore, arises the question of stability of the final images with respect to the windows, i.e., is it possible that starting from two spectrograms of the same signal for different windows the corresponding final images are very different from each other? The answer is:

Corollary 1. *Let $\varphi, \psi \in W^{1,\infty}(\mathbb{R})$ be real, symmetric and normalized windows and denote by u_0 and v_0 the corresponding spectrograms of a given signal $x \in L^2(\mathbb{R})$. Let u and v be the solutions of Problem P corresponding to the initial data u_0 and v_0, respectively. Then , for some constant $c > 0$,*

$$\sup_{0 \leq \tau \leq T} \|u(\tau, \cdot, \cdot) - v(\tau, \cdot, \cdot)\|_{L^\infty(\Omega)} \leq c\|\varphi - \psi\|_{L^2(\mathbb{R})}.$$

Proof. Let $\mathcal{G}_\eta x$ denote the Gabor's transform of x relative to the window η. The standard inequality $||a| - |b|| \leq |a - b|$ implies

$$\left||\mathcal{G}_\varphi x(\omega, t)|^2 - |\mathcal{G}_\psi x(\omega, t)|^2\right| \leq c_1 \left||\mathcal{G}_\varphi x(\omega, t)| - |\mathcal{G}_\psi x(\omega, t)|\right| \qquad (1.4)$$

$$\leq c_1 |\mathcal{G}_\varphi x(\omega, t) - \mathcal{G}_\psi x(\omega, t)|,$$

with $c_1 = ||\mathcal{G}_\varphi x(\omega, t)| + |\mathcal{G}_\psi x(\omega, t)||$. We have

$$|\mathcal{G}_\varphi x(\omega, t) - \mathcal{G}_\psi x(\omega, t)| \leq \int_{\mathbb{R}} |x(s)(\varphi(s - t) - \psi(s - t))e^{-i\omega s}|ds$$

$$\leq \|x\|_{L^2}\|\varphi - \psi\|_{L^2}. \qquad (1.5)$$

Taking the supremo in the left hand side of (1.4) and using (1.5) we obtain

$$\|u_0 - v_0\|_{L^\infty(\Omega)} \leq (\|u_0\|_{L^\infty(\Omega)}^{1/2} + \|v_0\|_{L^\infty(\Omega)}^{1/2})\|x\|_{L^2}\|\varphi - \psi\|_{L^2}. \qquad (1.6)$$

Finally, property (1.3) implies the result. □

Another stability question solved by Theorem 1 is whether the transformed spectrograms of two close signals relative to the same window are close or not. Since the proof is a trivial modification of the proof of Corollary 1, we omit it.

Corollary 2. *Let $x, y \in L^2(\mathbb{R})$ be two signals and $\varphi \in W^{1,\infty}(\mathbb{R})$ be a real, symmetric and normalized window. Let u_0 and v_0 be their spectrograms, and u and v be the corresponding solutions of Problem P. Then, for some $c > 0$,*

$$\sup_{0 \leq \tau \leq T} \|u(\tau, \cdot, \cdot) - v(\tau, \cdot, \cdot)\|_{L^\infty(\Omega)} \leq c\|x - y\|_{L^2(\mathbb{R})}.$$

For example, if x, n are signals, with n denoting a noise with unitary energy in L^2, and we define $y = x + \varepsilon n$, then, Corollary 2 implies that the corresponding spectrograms at *time* τ satisfy

$$\sup_{0 \leq \tau \leq T} \|u(\tau, \cdot, \cdot) - v(\tau, \cdot, \cdot)\|_{L^\infty(\Omega)} \leq c\varepsilon.$$

1.3 Numerical Experiments

In this section we present numerical demonstrations of the selective smoothing edge enhancement algorithm of [1] applied to the spectrograms of synthetic and field signals. We compare the results with the following methods: the 2D Stationary Wavelet Transform (2D-SWT) for 2D signal denoising, the Nonlinear Spectral Subtraction (NSS) based on [3] and the Stationary Wavelet Transform (1D-SWT) for 1D signal denoising, based on [10]. First, we describe the discrete PDE model.

1.3.1 PDE model discretization

We start by computing the spectrogram by applying the dfft to the convolution of the signal with the window. The dfft is evaluated in time intervals of size 2^w, with the *width*, w, usually in the range $8 - 12$. To obtain an image as *continuous* as possible, time intervals are overlapped according to the value of $p \in (0,1)$. In each of these intervals, we perform the convolution of the signal with a normalized discrete gaussian window with support on $(-t_0/2, t_0/2)$ and variance $\sigma = t_0/3$, where t_0 is the size of the time intervals (increasing with w).

Once the spectrogram is produced, it is normalized in the usual digital image range $[0, 255]$, obtaining in this way the initial datum for Problem P. We use a time explicit Euler scheme with finite differences in space to find the numerical approximation of the solution, u. We follow the discretization indicated in [1].

Summarizing, the parameters in the model come from three sources: the spectrogram definition, the image processing PDE and its numerical implementation. From the first we get the variance of the gaussian window, σ, which is determined by the width, w, and the overlapping, p. From the PDE we have the enhancement parameter, ϵ and the scale parameter, s. Finally, from the discretization we have the evolution step, $d\tau$, and the number of advances or iterations, k. In the experiments, we keep fixed those parameters which seems to be less sensible. More precisely, we always take

$$p = 0.99, \quad \epsilon = \frac{1}{2}\max|\nabla u_0|, \quad s = 1, \quad d\tau = 0.1.$$

Hence, the only parameters we play with in the experiments are w and k. Both are very related to the computer execution time since the width determines the time-frequency grid size. It is not clear how to fix them a priori. On one hand, the width is related to the smoothness of the discrete spectrogram and variations of this parameter may induce breaks in the lines of instantaneous frequencies, among other effects. Similarly, when the number of iterations increases the image gets more and more diffused making possible that some not very neatly defined edges may disappear.

1.3.2 Comparisons and results

To show the advantages of our technique, in the subsequent plots we used a simple algorithm to produce candidates to IF lines of the corresponding spectrograms. Let $\Omega = [0, T] \times [0, F]$ be the time-frequency domain of the image and $u : \Omega \to [0, 255]$ be the starting image. We consider its truncation

$$v(t, \omega) = \begin{cases} u(t, \omega) & \text{if } u(t, \omega) \geq \beta \\ 0 & \text{elsewhere,} \end{cases}$$

with $\beta = \text{Mean}_\Omega(u)$ in the experiments. For each $t \in [0, T]$, we consider the N connected components of the set $\{\omega \in (0, F) : v(t, \omega) > 0\}$, say $C_n(t)$, for $n = 1, \ldots, N(t)$, and define the function

$$\text{IF}(t, n) = \frac{\int_{C_n(t)} \omega v(t, \omega) d\omega}{\int_{C_n(t)} v(t, \omega) d\omega}, \tag{1.7}$$

which is the frequency gravity center of the component $C_n(t)$. In this way, we shrink each connected component to one point to which we assign the average image intensity through the function $\text{INT}(t, n) = \text{Mean}_{C_n(t)}(v(t, \cdot))$. Finally, we plot function IF only for components with averaged intensity, INT, greater than a certain threshold, $i \in [0, 255]$. This final image does not seem to be very sensible under small perturbations of the parameters β and i.

Experiment 1. We illustrate some noise reduction algorithms applied to a synthetic signal having in mind that our aim is to obtain a good spectrogram representation of the signal for later IF recognition, and not the 1D signal restoration. We used the following methods: our PDE based algorithm, the 1D and 2D Stationary Wavelet Transforms (SWT), see [10, 14], and the Nonlinear Spectral Substraction Method (NSS), see [3]. Among these, only the spectral based methods gave good results. The 1D SWT, based in high frequencies filtering of a multi-frequency model, do not produce, as it was expected, good spectrogram images for the processed signal. Therefore, we only show the results concerning the spectral based methods.

We used a one sec. 6KHz synthetic signal composed by the addition of two signals. The first is an addition of pure tones and chirps,

$$x_1(t) = c_1(\sin 2\pi 1000t + \sin 2\pi 1100t + \sin 2\pi 1300t^2 + \sin 2\pi 800t^3),$$

while the second, x_2, is a uniformly distributed real random variable. We normalize them to have unitary energy, $\|x_i\|_{L^2} = 1$ (so the constant c_1), and define the test signal as $x = x_1 + x_2$, i.e., with SNR $= 0$ dB. We fix the window width as $w = 10$ and perform $k = 50$ iterations of the PDE algorithm.

In Fig. 1.1, we show the spectrograms of the clean and noisy signals, x_1 and x, respectively, and the spectrogram resulting from the PDE algorithm. Notice that even for very close instantaneous frequency lines, the PDE processed spectrogram keeps them separated, despite being produced by a diffusive transformation of the noisy signal.

In Fig. 1.2, we plot the IF function defined by (1.7), obtained from the outputs of the PDE, the 2D-SWT and the NSS methods, for threshold levels

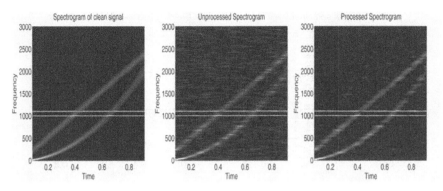

Fig. 1.1. Spectrograms of a synthetic signal with $SNR = 0$, (Experiment 1).

$i = 5$ and $i = 10$, respectively. We observe that both the PDE and the 2D-SWT perform much better than the NSS. It is also noticeable that the PDE method is less sensible to variations of the threshold level than the 2D-SWT.

We finally computed two objective 1D signal quality measures of the performance of the PDE and the 2D-SWT models: the SNR and the segmented SNR (for 200 frames). We obtained the following figures:

$$SNR_{PDE} = 3.04, \quad SGM - SNR_{PDE} = 3.75,$$
$$SNR_{SWT} = 2.12, \quad SGM - SNR_{SWT} = 2.87,$$

showing some better performance of the PDE model against the SWT model, in this example.

Experiment 2. We used a recording by [11], from where we extracted a signal of app. 0.55 sec. which is affected by a strong background noise. We set the window width $w = 10$, and performed $k = 200$ iterations. In the first row of Fig.1.3 we plot the spectrogram of the original signal (initial datum) and the processed spectrogram, and in the second row, the corresponding IF lines for the threshold value $i = 3$. We identify three possible howls, one with two harmonics in the approximated steady frequencies 400 and 800 Hz, another in about 600 Hz (decreasing in time), and finally, another with two harmonics starting at 1000 (decreasing) and 700 Hz, respectively, although the latter becomes too weak to be detected after a while. We notice the large qualitative difference between the IF lines detection of the noisy and the processed image, plotted in the second row.

1.4 Conclusions

This research establishes the first step of a new methodology for estimating wolves populations by analyzing their chorus field recorded signals with signal

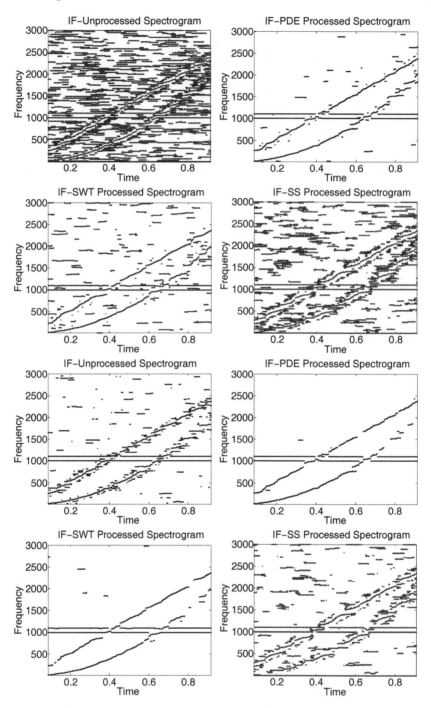

Fig. 1.2. IF lines corresponding to Experiment 1 for the unprocessed spectrogram and for the spectrogram processed by the PDE, 2D-SWT and NSS methods. First and second rows correspond to the threshold level $i = 5$. Second and third rows to $i = 10$.

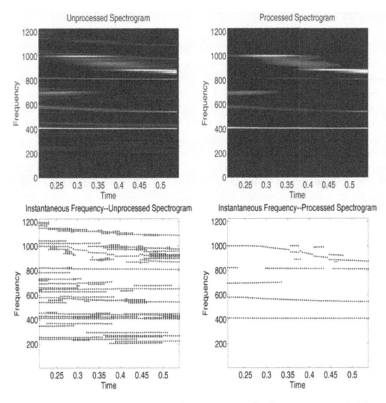

Fig. 1.3. Spectrograms and IF lines (Experiment 2) of a very noisy field recorded signal. The threshold level is taken as $i = 3$.

and image processing techniques, e.g., the noise reduction of the image representing a time-frequency distribution of the signal, such as the spectrogram. The second step, consisting on identifying the instantaneous frequency lines corresponding to each individual, is in progress.

A possible framework for the spectrogram image analysis, the use of partial differential equations based models, is investigated. We deduced stability results for the processed spectrogram with respect to perturbations due to noise or to changes of window functions. We demonstrated this technique on field recorded signals and artificial signals. For the latter, we compared the results with other well known techniques. These comparisons indicate, on one hand, that the use of image denoising wavelet based methods give similar results than the PDE method and it is, therefore, an alternative to take into account. On the other hand, denoising algorithms acting directly on the one-dimensional signal do not provide good spectrogram images for the second step of the method, the IF estimation.

References

1. Álvarez L, Lions PL, Morel JM (1992) SIAM J Numer Anal 29(3):845-866
2. Aubert G, Kornprobst K (2002) Mathematical problems in image processing. Springer, New York
3. Berouti M, Schwartz R, Makhoul J (1979) Proc IEEE Intl Conf Acoust Speech and Signal Proc: 208-211.
4. Chandra Sekhar S, Sreenivas TV (2003) Signal Process 83(7):1529–1543
5. Crandall MG, Ishii H, Lions PL (1992) Bull Amer Math Soc (NS) 27(1):1–67
6. Djurović I, Stanković L (2004) Signal Process 84(3):631–643
7. Dugnol B, Fernández C, Galiano G (2007) Appl Math Comput 186(1):820–830
8. Dugnol B, Fernández C, Galiano G, Velasco J (2007) Accepted in Appl Math Comput http://dx.doi.org/10.1016/j.amc.2007.03.086
9. Dugnol B, Fernández C, Galiano G, Velasco J (2007) In preparation
10. Donoho DL (1995) IEEE Trans on Inf Theory 41(3):613-627
11. Llaneza L, Palacios V (2003) Field recordings obtained in wilderness in Asturias, Spain
12. Mallat S (1998) A wavelet tour of signal processing. Academic Press, London
13. Mann S, Haykin S (1995) IEEE Trans. Signal Process 43(11):2745–2761
14. Matlab 7, The MathWorks, Inc.
15. Ozaktas HM, Zalevsky Z, Kutay MA (2001) The Fractional Fourier Transform with Applications in Optics and Signal Processing, Wiley, Chichester
16. Rabiner LR, Schafer RW (1988) Digital processing of speech signals. Prentice-Hall, New Jersey
17. Sapiro G (2001) Geometric partial differential equations and image analysis. Cambridge UP, Cambridge.
18. Skonhoft A (2006) Ecol Econ 58(4):830–841
19. Viswanath G, Sreenivas TV (2002) Signal Process 82(2):127–132
20. Zou H, Wang D, Zhang X, Li Y (2005) Signal Process 85(9):1813–1826

A Modified Mean Shift Algorithm For Efficient Document Image Restoration

Fadoua Drira, Frank Lebourgois, and Hubert Emptoz

SITIS 2006, Tunisia
{fadoua.drira,fank.lebourgeois,hubert.emptoz}@insa-lyon.fr

Summary. Previous formulations of the global Mean Shift clustering algorithm incorporate a global mode finding which requires a lot of computations making it extremely time-consuming. This paper focuses on reducing the computational cost in order to process large document images. We introduce thus a local-global Mean Shift based color image segmentation approach. It is a two-steps procedure carried out by updating and propagating cluster parameters using the mode seeking property of the global Mean Shift procedure. The first step consists in shifting each pixel in the image according to its *R-Nearest Neighbor Colors* (*R-NCC*) in the spatial domain. The second step process shifts only the previously extracted local modes according to the entire pixels of the image.

Our proposition has mainly three properties compared to the global Mean Shift clustering algorithm: 1) an adaptive strategy with the introduction of local constraints in each shifting process, 2) a combined feature space of both the color and the spatial information, 3) a lower computational cost by reducing the complexity. Assuming all these properties, our approach can be used for fast pre-processing of real old document images. Experimental results show its desired ability for image restoration; mainly for ink bleed-through removal, specific document image degradation.

Key words: Mean Shift, segmentation, document image, restoration, ink bleed-through removal.

2.1 Introduction

Image segmentation techniques play an important role in most image analysis systems. One of their major challenge is the autonomous definition of color cluster number. Most of the works require an initial guess for the location or the number of the colors or clusters. They have often unreliable results since the employed techniques rely upon the correct choice of this number. If it is correctly selected, good clustering result can be achieved; otherwise, image segmentation cannot be performed appropriately. The Mean Shift algorithm, originally advanced by Fukunaga [1], is a general nonparametric clustering technique. It does not require an explicitly definition of the cluster number. This number is obtained automatically.It is equal to the number of the extracted centers of the multivariate distribution underlying the feature space. Advantages of feature space methods are the global representation of the original data and the excellent tolerance to noise. This property is a robust process for degraded document images that legibility is often compromised due to the presence of artefacts in the background [2]. Processing of such degraded documents could be of a great benefit, especially to improve human readability and allow further application of image processing techniques. Under its original implementation, the global Mean Shift algorithm cannot be applied on document images. In fact, documents are generally digitized using high resolution, which provides large digital images that slow down the segmentation process. Therefore, with the increase of the pixel numbers in the image, finding the closest neighbors of a point in the color space becomes more expensive. In this paper, we propose an improved Mean Shift based two-steps clustering algorithm. It takes into account a constrained combined feature space of the both color and spatial information. In the first step, we shift each pixel in the image to a local mode by using the *R-Nearest Neighbor Colors* in the spatial domain. These neighbors are extracted from an adaptive sliding window centred upon each pixel in the image. R represents an arbitrary predefined parameter. In the second step, we shift ,using all pixels,the previously extracted local modes to global modes. The output of this step is a collection of global modes. These modes are candidate cluster centers.

This paper is organized as follows. Section 2 describes briefly the global Mean Shift clustering algorithm using the steepest ascent method. The proposed algorithm with local constrained Mean Shift analysis is introduced and analyzed in Section 3. Experimental segmentation results, using our proposition for degraded document image restoration and more precisely for ink bleed-through removal, are presented in section 4.

2.2 The global Mean Shift: Overview

Before treating the proposed algorithm based on a local-global Mean Shift procedure, we would explain the global Mean Shift and its related clustering algorithm in brief [3]. For a given image with N pixels, we use x_i to denote the observation at the ith color pixel. $\{x_i\}_{i=1\cdots N}$ is an arbitrary set of points defined in the R^d d-dimensional space and k the profile of a kernel K such that:

$$K(x) = c_{k,d} \, k(\|x\|^2) \, . \tag{2.1}$$

The multivariate kernel density estimator, with kernel $K(x)$and window radius (bandwidth) h is given by:

$$\widehat{f}(x) = \frac{c_{k,d}}{nh^d} \sum_{i=1}^{N} k\left(\left\|\frac{x - x_i}{h}\right\|^2\right) \, . \tag{2.2}$$

Although other kernels could be employed, we restrict this Mean Shift study to the case of the uniform kernel. The standard Mean Shift algorithm is defined as steepest gradient ascend search for the maxima of a density function. It requires an estimation of the density gradient using a nonparametric density estimator [3]. It operates by iteratively shifting a fixed size window to the nearest stationary point along the gradient directions of the estimated density function

$$\nabla \widehat{f_{h,k}}(x) = \frac{2c_{k,d}}{nh^d} \sum_{i=1}^{N} \nabla k \left(\left\| \frac{x - x_i}{h} \right\|^2 \right)$$

$$= \frac{c_{k,d}}{nh^{d+2}} \sum_{i=1}^{N} g \left(\left\| \frac{x - x_i}{h} \right\|^2 \right) \left[\frac{\sum_{i=1}^{N} x_i g(\left\| \frac{x - x_i}{h} \right\|^2)}{\sum_{i=1}^{N} g(\left\| \frac{x - x_i}{h} \right\|^2)} - x \right]. \quad (2.3)$$

We denoted

$$g(x) = -k'(x) \quad (2.4)$$

which can in turn be used as profile to define a kernel $G(x)$ where

$$G(x) = c_{g,d} \ g(\|x\|^2). \quad (2.5)$$

The kernel $K(x)$ is called the shadow of $G(x)$ [3]. The last term (6)

$$m_{h,G}(x) = \frac{\sum_{i=1}^{N} x_i g(\left\| \frac{x - x_i}{h} \right\|^2)}{\sum_{i=1}^{N} g(\left\| \frac{x - x_i}{h} \right\|^2)} - x \quad (2.6)$$

shows the Mean Shift vector equal to the difference between the local mean and the center of the window. One characteristic of this vector, it always points towards the direction of the maximum increase in the density. The converged centers correspond to the modes or the centers of the regions of high data concentration. Figure 2.1 illustrates the principle of the method. The window tracks signify the steepest ascent directions. The mean shift vector, proportional to the normalized density gradient, always points toward the steepest ascent direction of the density function. It can be deducted that searching the modes of the density is performed by searching the convergent points of the mean shift without estimating the density [3].

The global Mean Shift clustering algorithm can be described as follows:

1. Choose the radius of the search window,
2. Initialize the location of the window x^j, $j = 1$,
3. Compute the Mean Shift vector $m_{h,G}(x^j)$,
4. Translate the search window by computing $x^{j+1} = x^j + m_{h,G}(x^j)$, $j = j + 1$,
5. Step 3 and step 4 are repeated until reaching the stationary point which is the candidate cluster center.

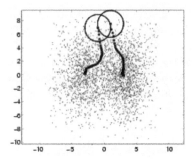

Fig. 2.1. Mean Shift mode finding: Sucessive computations of the Mean Shift define a path to a local density maximum

2.3 A local-global Mean Shift algorithm

2.3.1 The proposed local Mean Shift

The global Mean Shiftalgorithm, under its original form, defines a neighborhood around the current point in the feature space related to the color information. The neighborhood refers to all the pixels contained in the sphere of a given arbitrary radius σ_R centred on the current pixel. It is extracted from a fixed size window and used for the *Parzen* window density estimation. Applying Mean Shift leads to find centroids of this set of data pixels. The proposed Mean Shift algorithm called the local Mean Shift algorithm is an improved version of the global Mean shift algorithm by reducing its complexity. Our main contribution consists in introducing a constrained combined feature space of the both color and spatial information. Constraints are mainly introduced in the definition of a neighborhood necessary for the estimation of the Mean Shift vector. Therefore, we introduce the concept of a new neighborhood defined by the *R-Nearest Neighbor Colors* . It represents the set of the R nearest colors in the spatial domain extracted from an adaptive sliding window centred upon each studied data pixel in the image. R is an arbitrary predefined parameter. More precisely, we define the $R\text{-}NNC(X)$ the R spatially nearest points from a given pixel X and having a color distance related to X less than σ_R.The studied neighborhood of each pixel in the image, originally detected in a fixed window width, is modified in order to be defined from a gradually increasing window size. Starting from a 3x3 window size centred on a given data pixel X, we set for each neighbor Y within the window its color distance from X. Then, we record all the neighbors having a color distance less than an arbitrary fixed value σ_R. If the number of the memorized data pixels is less than a fixed arbitrary value R, we increase the size of the window. We iterate the process of neighbors' extraction and window increasing while the desired number of neighbor's or the limit size of the window is not reached. The selection of the neighbors is as follow:

$$R - NNC(X) = \left\{ \begin{array}{c} Y/d_{color}(X,Y) < \sigma_R \text{ is the spatially} \\ \text{nearest neighbor of } X \end{array} \right\}$$

Intuitively, using here a progressive window size is of beneficial. This comes from the fact that computation of the mode is restricted inside a local window centred on a given data pixel and more precisely restricted on the colorimetrically and spatially nearest neighbors. By doing so, we guarantee an accurate convergence of the Mean Shift in few iterations. Figure 2.2 illustrates an example of the Mean Shift vector direction that points towards the direction of the most populated area. Furthermore, it is evident that the local mode closest to the value of the central pixel is a far better estimate of the true value than the average of all color values.

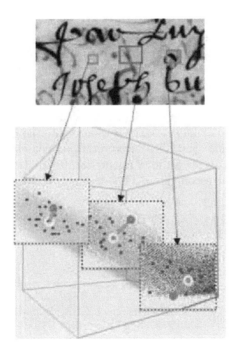

Fig. 2.2. Scan of a manuscript and a zoom on a located window in the $L^*u^*v^*$ cube after local Mean Shift application. Blue points are the R neighbors; red circle is a studied data image pixel; yellow circle is the extracted local mode.

2.3.2 The proposed segmentation algorithm

The proposed segmentation algorithm follows the steps as below:

1. Run the local Mean Shift algorithm starting from each pixel X of the data set (converted to the feature space $L^*u^*v^*$) and shifting over the

R-NNC(X) neighborhood. Once all the data pixels are treated, different local maxima of pixel densities are extracted.

2. Run the global Mean Shift algorithm starting from the extracted local modes and shifting over all pixels of the data image to reach the global maxima.

3. Assign to all the pixels within the image the closest previously extracted mode based on their color distance from each mode. The number of significant clusters present in the feature space is automatically established by the number of significant detectedmodes. Therefore, the global extracted modes can be used to form clusters.

Based on the above steps, it is clear that the first one generates an initial over-segmentation. This can be considered as a good starting point for the second step which is important to find the global modes. In fact, the over-segmentation is absolutely related to an important number of the local extracted modes. This number depends mainly on the R predefined value. If the value of R increases, the number of the extracted modes decreases. Consequently, choosing small values reduce neighbor's number related to each given data image pixel. Hence, we generate an important number of the extracted local modes after the first step. Figure 2.3 illustrates in the first three instances the distribution of the extracted local modes for different value of R. All these values are given as an example and they change enormously from one image to another. Nevertheless, the given interpretation remains the same. The last instance in figure 2.3 gives an idea about the distribution of the extracted global modes after the second step. This result is obtained for $R=25$.

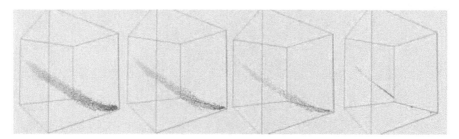

Fig. 2.3. From left to right: the three first figures correspond to the distribution of The K extracted local modes for different R values:$R=25$, $K=870$; $R=100$, $K=431$; $R=400$, $K=236$ repectively ;The last figure is related to the distribution of the N global modes for $R=25$ given as an example

2.3.3 Complexity estimation

The application of the local Mean Shift as a first step has a strong impact on the computational time as well as on the quality of the final result. This step is important to provides efficient starting points for the second step. These points are sufficiently good local maxima. Therefore, finding global modes, which is the aim of the second step, will be performed with a reduced complexity. The final results, as it will be illustrated in the next section, are more likely to be satisfactory. Without optimisation, the computational cost of an iteration of the global Mean Shift is $O(N \times N)$, where N is the number of image pixels. The most expensive operation of the global Mean Shift algorithm is finding the closest neighbors of a point in the color space. Using the most popular structures, the KD-tree, the complexity is reduced to $O(N \log N)$ operations, where the proportionality constant increases with the the space dimension. Obviously, our proposition reduces this time complexity, in the ideal case, to $O(N \times (R+M))$, where R is the number of the spatially and colorimetically nearest neighbors and M the number of the extracted local modes after the first step. The value of M depends on the content of the processed images. Therefore, we are unable to estimate in advance the computational time.

2.3.4 Performance comparison

The proposed local-global Mean Shift clustering algorithm is an improved version of the global Mean Shift. Moreover, our proposition takes benefit from a combined feature space that consists in a combination of the spatial coordinates and the color space. Such space has been already proposed in the literature as a modified Mean Shift based algorithm, we note it here as the spatial Mean Shift [4]. The main difference between these three procedures is correlated to the neighborhood of each data pixel. We note $N\_global\_MS(X)$, $N\_spatial\_MS(X)$and $N\_local\_MS(X)$ the studied neighborhood related respectively to the spatial, global and local-global Mean Shift in a first step iteration and for a given data pixel X.

$$N\_global\_MS(X) = \{Y/d_{color}(X,Y) < \sigma_R\}$$
$$N\_spatial\_MS(X) = \{Y/d_{color}(X,Y) < \sigma_R \text{ and } d_{spatial}(X,Y) < \sigma_S\}$$
$$N\_local\_MS(X) = \{R\text{-}NNC(X)\}$$

For instance, the $N\_global\_MS(X)$ involves all data pixels in the image having a color distance from X less than σ_R, a fixed size window. The $N\_spatial\_MS(X)$ represents the set of neighbors having a color distance from X less than σ_R and located in a distance less than σ_S in the spatial domain. Compared to these two procedures, if their studied neighborhood is detected in a fixed window width including the color information in the global Mean Shift and the both of color and spatial information in the spatial Mean Shift, the local Mean Shift is not restricted to a fixed window size. It depends on the total number of spatial neighbors having a color distance less than σ_R. Therefore, the $N\_local\_MS(X)$ is defined from a gradually increasing window size until reaching a predefined number of neighbors. If the global Mean Shift algorithm is a time-consuming process, the spatial Mean Shift achieves a low computational cost with efficient final results for image segmentation. One question, could be evoqued here, why defining a local-global Mean Shift algorithm if we already have an efficient improved Mean Shift with lower complexity? In fact, the segmented image with the spatial Mean Shift is generally over-segmented to a great number of small regions. Some of them must be finally merged by using heuristics. In the case of document images, the spatial Mean Shift clustering algorithm is not efficient since it breaks the strokes of the handwritten foreground and over-segments the background. Moreover, the major challenge of this Mean Shift variant is the adaptive specification of the two window widths according to the both of data statistics and color domains in the image. These two parameters are critical in controlling the scale of the segmentation result. Too large values result in loss of important details, or under-segmentation; while too small values result in meaningless boundaries and excessive number of regions, or over-segmentation. It is obviously that our proposition is different from the spatial Mean Shift clustering algorithm as it is a two-steps algorithm. The advantage of using the local Mean Shift followed by the global Mean Shift rather than the direct use of the spatial Mean Shift is twofold. First, we can omit the use of statistics to merge regions detected after a spatial Mean Shift application in order to have significant parts. Second, we guarantee to generate a sufficient neighbor's number necessary in the shifting process. Figure 2.4 illustrates the final result obtained after the three procedures application on an extract of a document image.

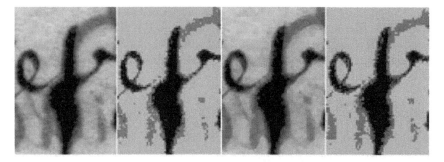

Fig. 2.4. From left to right: an extract of a bleed-through degraded document, the segmented image with the global Mean Shift, the segmented image with the spatial Mean shift and the segmented image with the local-global Mean Shift

2.4 Experimental results: Segmentation for document image restoration

2.4.1 Problem statement

Image segmentation and denoising are two related topics and represent fundamental problems of computer vision. The goal of denoising is to remove noise and/or spurious details from a given corrupted digital picture while keeping essential features such as edges. The goal of segmentation is to divide the given image into regions that belong to distinct objects. For instance, our previous work [2] proposes such technique application as a solution for the removal of ink bleed-through, a specific degradation for document images. This degradation is due to the paper porosity, the chemical quality of the ink, or the conditions of digitalization. The result is that characters from the reverse side appear as noise on the front side. This can deteriorate the legibility of the document if the interference acts in a significant way. To restore these degraded document images, this noise is simulated by new layers at different gray levels that are superposed to the original document image. Separating these different layers to improve readability could be done through segmentation/classification techniques. In a first study, we tested the performance of the most popular algorithm among the clustering ones, the K-means, known for its simplicity and efficiency. Nevertheless, this technique remains insufficient for restoring too degraded document images. Indeed, ink bleed-through removal could be considered as a three-class segmentation problem as our aim consists in classifying pixel document images into (1) background, (2) original text, and (3) interfering text. According to this hypothesis, a K-means ($K{=}3$) might be sufficient to correctly extract the text of the front side. But this is not the case (Fig.2.5).

Fig. 2.5. Results of the *3*-means classification algorithm on a degraded document image

Intuitively, other variants of the K-means clustering algorithm are employed to resolve this problem. In this study, we will focus on techniques based on the extension of K-means. For a complete sate-of-the-art ink bleed-through removal techniques, please refer to our previous work [2]. One variant based on a serialization of the K-means algorithm consists in applying sequentially this algorithm by using a sliding window over the image [5]. This process leads to an automatic adjust of the clusters during the windows displacement, very useful for a better adaptation to any local color modification. This approach gives good results but it is a supervised one as the choice of some parameters such as the number of clusters and the color samples for each class are not done automatically. We reveal this problem mainly when the text of the front side has more than one color. This problem remains also problematic to another variant of the K-means algorithm applied on degraded document images. This variant [6] consists in a K-means (K=2) recursive application on the decorrelated data with the Principal Component Analysis (PCA). It generates a binary tree that only the leaves images satisfying a certain condition on their logarithmic histogram are processed. The definition of the number of classes is avoided here and the obtained results justify the efficiency of this approach. Nevertheless, for a document image having more than one color on the text of its front side, a certain number of leaves images corresponding to the number of colors used in the front text must be combined. In this case, the choice of these different leaves cannot be done automatically and the intervention of the user is obviously necessary.

Consequently, the accuracy of such techniques related to the accuracy of K-means clustering results is inevitably compromised by 1) the prior knowledge of the number of clusters and 2) the initialisation of the different centers generally done randomly. The K-means clustering can return erroneous results when the embedded assumptions are not satisfied. Resorting to an approach which is not subject to these kind of limitations will certainly leads to more accurate and robust results in practice. Moreover, ink bleed-through generates random features that only powerful flexible segmentation algorithm could cope with it. Intuitively, according to our study, we have noticed the flexibility of a statistical data based segmentation algorithm which can accurately classify random data points into groups. One of the most promising techniques of this category is the Mean Shift which represents the core technique of our proposition; the local-global Mean Shift algorithm.

2.4.2 Performance evaluation

Experiments were carried out to evaluate the performance of our approach based on a modified Mean Shift algorithm. For our simulations, we set σ_R, the minimum color distance between a starting point and its neighbor, to the value of 6 and the number R of the extracted neighbors to the value of 25. Results of applying the proposed approach on degraded document images are displayed in the figure 2.6. These documents, which have been subject to ink bleed-through degradation, contain the content of the original side combined with the content of the reverse side. These images are first mapped into the $L*u*v*$ feature space. This color space was employed since its metric is a satisfactory approximation to Euclidean distance. Then, we apply our algorithm to form clusters. The images resulting from the application of our approach on the degraded document images, shown in the figure 2.6, are correctly restored. We clearly notice, compared with the test images, that the interfering text has been successfully removed. Moreover, the segmentation obtained by this technique looks as similar as or better than that obtained by the global Mean Shift (Fig.2.4). The important improvement is noticed with a significant speedup. This is due to the selective processing of the data image pixels ; only the R nearest color neighbors to a given pixel are processed. By modifying the global Mean Shift algorithm, we reduce the number of iterations necessary for finding the different modes and thus to achieve convergence. In fact, the processing of a 667X479 color document image with $R=25$ and $\sigma_R=6$, is done in 470 seconds with our proposition and in approximately 19 hours with the global Mean Shift algorithm. The first step of our method generates 1843 local modes and takes 70 seconds. The second step, consisting in shifting these modes according to all data pixels takes 400 seconds. For the global Mean Shift algorithm, we have 319493 pixels to shift according to all data pixels. This clearly explains the high computational cost time. These different values are related to the second horizontal original color image of the figure 2.6.

2.5 Conclusion

We have presented in this study an improvement of the global Mean Shift algorithm in order to reduce its computational cost and thus making it more flexible for large document image processing. Our proposition, called the local-global Mean Shift clustering algorithm, has been successfully applied for document image restoration, more precisely for ink bleed-through removal. This algorithm is validated with good results on degraded document images. Our goal was to produce an algorithm that retains the advantages of the global Mean Shift algorithm but runs faster. This is correctly achieved. Nevertheless, the performance of our proposition is dependent on the minimum distance that must be verified between a given pixel and its neighbor that it will be included in the first shifting process. This distance is defined the same in the different steps of the algorithm. In this context, the local-global Mean Shift algorithm could be a subject of ameliorations. For instance, this color distance could vary from one iteration to another. This could be based on predifined contraints. Varying this number could add an adaptative strategy with better results. Subsequent investigations in not applying the Mean Shift procedure to the pixels which are on the mean shift trajectory of another (already processed) pixel could also be done. Our future research will investigate all these different ideas and test the proposed method on a large set of document images.

References

1. K. Fukunaga, Introduction to Statistical Pattern Recognition. Boston, MA: Academic Press, 1990.
2. F. Drira, Towards restoring historic documents degraded over time. Dans Second IEEE International Conference on Document Image Analysis for Libraries (DIAL2006), Lyon, France. pp. 350-357. ISBN 0-7695-2531-4. 2006.
3. Y. Cheng, Mean shift, mode seeking, and clustering. Pattern Analysis and Machine Intelligence, IEEE Transactions, vol. 17, pp. 790 . 799, 1995.
4. D. Comaniciu and P. Meer, Mean shift: A robust approach toward feature space analysis. IEEE Transactions on Pattern Analysis and Machine Intelligence, vol. 24, no. 5, pp. 603.619, 2002.
5. Y. Leydier, F. LeBourgeois, H. Emptoz, Serialized k-means for adaptive color image segmentation . application to document images and others. DAS 2004, LNCS 3163, Italy, September 2004, 252-263.
6. F. Drira, F. Lebourgeois, H. Emptoz Restoring Ink Bleed- Through Degraded Document Images Using a Recursive Unsupervised Classification Technique. DAS2006, LNCS 3872. Nelson, New Zealand, 2006, 38-49.

Fig. 2.6. Original bleed-through degraded document images and their restored version with our proposed local-global Mean Shift algorithm

3

An Efficient Closed Form Approach to the Evaluation of the Probability of False Alarm of the ML-CFAR Detector in a Pulse-to-Pulse Correlated Clutter

Toufik Laroussi[1] and Mourad Barkat[2]

[1] Laboratoire Signaux et Systèmes de Communication, Département
d'Electronique, Université de Constantine, Constantine 25010, Algeria.
toufik_laroussi@yahoo.fr
[2] Department of Electrical Engineering, American University of Sharjah, P. O.
Box 26666, Sharjah, United Arab Emirates. mbarkat@aus.edu

Summary. The problem of adaptive CFAR regulation in a pulse-to-pulse partially Rayleigh correlated clutter is addressed. The clutter is modeled as a first-order Markov Gaussian process and its covariance matrix is assumed to be known. The theorem of residues is used to derive an exact expression for the probability of false alarm P_{fa} for the mean level (ML) detector. We show that it depends on the degree of pulse-to-pulse correlation of the clutter samples.

Key words: CFAR detection, pulse-to-pulse correlated clutter, residues.

3.1 Introduction

From experimental data, clutter is modeled by either Rayleigh, log-normal, Weibull, or K distribution [1-3]. Dealing with either single pulse or multiple pulses or even distributed architectures with correlated clutter samples models, aims for improved detection while maintaining a constant false alarm rate have led to consideration of several types of adaptive CFAR detectors.

The problem of partially correlated clutter has been treated in [4-7]. In [4] Farina et al considered the problem of detecting pulse-to-pulse partially correlated target returns in pulse-to-pulse partially correlated clutter returns. They suggested a batch detector and a recursive detector to estimate the clutter covariance matrix. They showed, by Monte Carlo simulations, that the threshold multiplier is independent of the degree of pulse-to-pulse correlation among the clutter samples for a number of range cells around ten. In [5], Himonas et al proposed the Generalized GCA-CFAR detector that adapts

not only to changes in the clutter level but also to changes in the clutter covariance matrix, i.e., they considered the case where the correlated statistics
of the received clutter returns are unknown. In their analysis, they assumed
a single pulse, processing a spatially correlated clutter whereas in the ML-
CFAR detector, we treat multiple pulses, processing a pulse-to-pulse partially
correlated Rayleigh clutter. They also assumed Markov's type homogeneous
clutter of much higher power than thermal noise. Al Hussaini et al studied in
[6] the detection performance of the CA-CFAR and OS-CFAR detectors considering a non conventional time diversity technique. Their schemes processed
correlated clutter in the presence of interfering targets. In [7], El Mashade
introduced the case where the use of the moving target indication (MTI) filter output introduces a correlated clutter even though if its input signal is
uncorrelated. He analyzed the CA-CFAR family for the multiple target environments scenario.

In summary, the work listed above did not show the effect of the pulse-
to-pulse partially Rayleigh correlated clutter on the analytical expression of
the P_{fa}. All of this, has led us to introduce, in this chapter, a mathematical
model to represent the general case of processing a pulse-to-pulse partially
Rayleigh correlated clutter and Rayleigh but uncorrelated thermal noise. The
chapter is organized as follows. In Section 3.2, we formulate the statistical
model and set the assumptions. Then, in Section 3.3, we investigate the effect
of the temporal correlation of the clutter on the probability of false alarm
and particularly on the CFAR parameter T for the ML detector employing
M-pulse noncoherent integration. Next, in Section 3.4, we show, by means
of computer simulation, the effect of the degree of correlation of the clutter
returns and the number of processed pulses M, on the threshold multiplier
T and on the probability of false alarm as well. As a conclusion, Section 3.5
summarizes the results of this contribution.

3.2 Statistical Model

The received signal $r(t)$ is processed by the in-phase and quadrature phase
channels. Assuming a target embedded in correlated noise, the in-phase and
quadrature phase samples $\{u_{ij}\}$ and $\{v_{ij}\}$ at pulse i and range cell j, respectively, $i = 1, 2, 3, \ldots, M$ and $j = 1, 2, 3, \ldots, N$, are observations from
Gaussian random variables. M and N are the number of radar processed
pulses and the number of reference range cells, respectively. Assuming that
the total clutter plus noise power is normalized to unity, the integrated output
of the square law detector is given by

$$q_j = \frac{1}{2} \sum_{i=1}^{M} \left\{ u_{ij}^2 + v_{ij}^2 \right\} \quad j = 1, 2, \ldots, N \qquad (3.1)$$

where

$$u_{ij} = a_{ij} + c_{ij} \quad i = 1, 2, \ldots, N \quad \text{and} \quad j = 1, 2, \ldots, N \qquad (3.2)$$

$$v_{ij} = b_{ij} + d_{ij} \quad i = 1, 2, \ldots, N \quad \text{and} \quad j = 1, 2, \ldots, N \tag{3.3}$$

The a_{ij}, c_{ij}, b_{ij} and d_{ij} represent the in-phase and quadrature phase samples of the correlated clutter and the thermal noise, respectively. The clutter samples are assumed to be first-order Markov processes with zero mean and variance σ_c^2 and are identically distributed but correlated (IDC) from pulse-to-pulse and uncorrelated from cell-to-cell. The thermal noise samples are assumed to be independent and identically distributed (IID) random variables with zero mean and variance σ_n^2 from pulse-to-pulse and from cell-to-cell. The co-variance matrices of the clutter and noise processes are denoted $\mathbf{\Lambda_c}$ and $\mathbf{\Lambda_n}$, respectively.

Let the overall clutter plus noise variance be $\sigma_{cn}^2 = \sigma_n^2 + \sigma_c^2$ and define the clutter-to-noise ratio as $CNR = \frac{\sigma_c^2}{\sigma_n^2}$. Thus, the $(i, j)^{th}$ element of the pulse-to-pulse covariance matrix $\mathbf{\Lambda_{cn}}$, can be shown to be [5]

$$[\mathbf{\Lambda_{cn}}]_{i,j} = \begin{cases} \sigma_{cn}^2 \left(\frac{\delta_{ij} + CNR}{1 + CNR} \right) \rho_c^{|i-j|} & 0 < \rho_c < 1 \\[2mm] \sigma_{cn}^2 \, \delta_{ij} & \rho_c = 0 \\[2mm] \sigma_{cn}^2 \left(\frac{\delta_{ij} + CNR}{1 + CNR} \right) & \rho_c = 1 \end{cases} \tag{3.4}$$

Note that this correlation model is exponentially decaying as a function of time difference $|i - j|$. $\rho_c \triangleq exp\left(-T\omega_c\right)$, is the correlation coefficient between pulse-to-pulse received clutter samples, T is the pulse repetition interval (PRI) and $f_c \triangleq \frac{\omega_c}{2\pi}$ is the mean Doppler frequency of the clutter signal [5]. δ_{ij} is the Kronecker delta. For convenience, we assume that the clutter is stationary and that the PRI is constant.

In the absence of a target, the detection performance is based upon the statistics of q, which is given by

$$q = \frac{1}{2} \sum_{i=1}^{M} \{u_i^2 + v_i^2\} \tag{3.5}$$

where

$$u_i = a_i + c_i \quad i = 1, 2, \ldots, M \tag{3.6}$$

$$v_i = b_i + d_i \quad i = 1, 2, \ldots, M \tag{3.7}$$

The a_i, c_i, b_i and d_i represent the in-phase and quadrature phase samples of the correlated clutter and the thermal noise, respectively. In vector form, by accommodating $M \times 1$ vectors, we have

$$q = \frac{1}{2} \left\{ |\mathbf{U}|^2 + |\mathbf{V}|^2 \right\} \tag{3.8}$$

where

$$U = A + C \tag{3.9}$$

$$V = B + D \tag{3.10}$$

The system under consideration assumes that the random vectors A and B are IID stationary Gaussian and that the noise signal is additive. The test cell q is then compared to the adaptive threshold TQ to make a decision H_0, according to the following hypothesis test

$$q \underset{H_0}{\gtrless} TQ \tag{3.11}$$

where Q denotes the estimated background level and H_0 denotes the absence of a target.

3.3 Evaluation of the False Alarm Probability

According to equation (3.11), the probability of false alarm of a CFAR detector can be obtained by using the contour integral, which can also be expressed in terms of the residue theorem as [8, 9]

$$P_{fa} = -\sum_{i_0} res\left[s^{-1}\Phi_{q|H_0}(s)\Phi_Q(-Ts), s_{i_0}\right] \tag{3.12}$$

where res [.] denotes the residue. s_{i_0} ($i_0 = 1, 2, \ldots$) are the poles of the moment generating function (mgf) $\Phi_{q|H_0}(s)$, from a noise background, lying in the left-hand of the complex s-plane. $\Phi_Q(-Ts)$ is the mgf of the estimated background level at $s = -Ts$.

To simplify the analysis, we confine our attention, in the remainder of this chapter, to the case where the clutter power is much higher than the thermal noise power, i.e., the overall covariance matrix Λ_{cn} given by equation (3.4) is approximated by the clutter covariance matrix Λ_c and $\sigma_{cn}^2 = \sigma_c^2$.

In order to derive an expression for the P_{fa}, we must evaluate the mgf $\Phi_{q|H_0}(s)$ of the test cell q in the absence of the target and the mgf $\Phi_Q(-Ts)$ of the background level. The block diagram of the ML-CFAR detector integrating M pulses is given by Fig. 28.1.

3.3.1 Mgfs of the Test Statistic and the Background Noise

The mgf of q in the absence of the target can be obtained as [5]

$$\Phi_{q|H_0}(s) = \int_{-\infty}^{+\infty} p(q|H_0) \, exp(-sq) \, dq \tag{3.13}$$

where $p(q|H_0)$ is the probability density function (pdf) of the test cell q under hypothesis H_0. Note that $\Phi_{q|H_0}(s)$ is the Laplace transform of $p(q|H_0)$. Substituting equation(3.8) for q, we can write

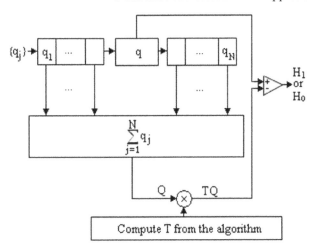

Fig. 3.1. ML-CFAR detector

$$\Phi_{q|H_0}(s) = \int_{-\infty}^{+\infty} \int_{-\infty}^{+\infty} p(\mathbf{U}, \mathbf{V}) \exp\left[-\frac{1}{2} s\left(|\mathbf{U}|^2 + |\mathbf{V}|^2\right)\right] d\mathbf{U} \, d\mathbf{V} \quad (3.14)$$

Assuming that the clutter power is much higher than the thermal noise power, we can write in terms of the clutter vectors

$$\Phi_{q|H_0}(s) = \int_{-\infty}^{+\infty} \int_{-\infty}^{+\infty} p(\mathbf{A}, \mathbf{B}) \exp\left[-\frac{1}{2} s\left(|\mathbf{A}|^2 + |\mathbf{B}|^2\right)\right] d\mathbf{A} \, d\mathbf{B} \quad (3.15)$$

Since the in-phase and quadrature phase samples of the clutter are IID, the joint pdf $p(\mathbf{A}, \mathbf{B})$ may be written as $p(\mathbf{A}, \mathbf{B}) = p(\mathbf{A})p(\mathbf{B})$ and $p(\mathbf{A}) = p(\mathbf{B})$. $p(\mathbf{A})$ is the multivariate Gaussian pdf with zero mean and covariance matrix $\Lambda_{\mathbf{c}}$ which gives rise to partially correlated clutter signals [10]. Thus

$$p(\mathbf{A}) = \frac{1}{(2\pi)^{\frac{M}{2}} |\Lambda_{\mathbf{c}}|^{\frac{1}{2}}} \exp\left(-\frac{1}{2} \mathbf{A}^T \Lambda_{\mathbf{c}}^{-1} \mathbf{A}\right) \quad (3.16)$$

where $|.|$ denotes the determinant Note that when $\rho_c \to 1$ the clutter samples are completely correlated. This corresponds to a singular covariance matrix Λ_c and equation (3.16) becomes [5]

$$p(\mathbf{A}) = \frac{1}{(2\pi)^{\frac{1}{2}}} \exp\left(-\frac{1}{2} a_1^2\right) \prod_{i=1}^{M} \delta(a_i - a_1) \quad (3.17)$$

where a_i (i=1, 2, ..., M) are the components of the in-phase clutter vector. Combining equations (3.15) and (3.16) and after some simple mathematical manipulations, the mgf of the cell under test under hypothesis H_0 can be expressed as

$$\Phi_{q|H_0}(s) = \frac{1}{|\mathbf{I} + \mathbf{\Lambda_c}\,s|} \left\{ \int_{-\infty}^{+\infty} \frac{\exp\left[-\frac{1}{2}\mathbf{A}^T\left(\mathbf{\Lambda_c}^{-1} + \mathbf{I}\,s\right)\mathbf{A}\right]}{(2\pi)^{M\!/_2}\left|\left(\mathbf{\Lambda_c}^{-1} + \mathbf{I}\,s\right)^{-1}\right|^{1\!/_2}}\,d\mathbf{A} \right\}^2 \quad (3.18)$$

where $\mathbf{I}$ is the identity matrix. Since the integral equals unity, the mgf of q under hypothesis H_0 is

$$\Phi_{q|H_0}(s) = \frac{1}{|\mathbf{I} + \mathbf{\Lambda_c}\,s|} \quad (3.19)$$

If the clutter signal arises from stationary process, then $\mathbf{\Lambda_c}$ is a symmetric Toeplitz matrix with M distinct positive real eigenvalues denoted by β_i, $i = 1, 2, \ldots, M$. Therefore, the determinant of equation(3.19) may be expressed as the product of its eigenvalues. Also, the background noise level Q is estimated by the average of the N reference cells ($Q = \sum_{n=1}^{N} q_n$). Since the q_n, $n = 1, 2, \ldots, N$, given by (1), are assumed to be IID and that we deal with uniform clutter power, then the mgfs of the cell under test under hypothesis H_0 and the ML detector can be expressed, respectively, as

$$\Phi_{q/H_0}(s) = \frac{1}{\prod\limits_{n=1}^{M}(1 + \beta_n s)} \quad (3.20)$$

and

$$\Phi_Q(s) = \frac{1}{\prod\limits_{n=1}^{M}(1 + \beta_n s)^N} \quad (3.21)$$

3.3.2 Analysis of the Probability of False Alarm

Since the probability of false alarm P_{fa} of the ML-CFAR detector is a function of the pulse-to-pulse clutter correlation coefficient ρ_c, we derive expressions of the P_{fa} for $\rho_c = 0$, $\rho_c = 1$ and $0\langle\rho_c\langle 1$.

Case 1. $\rho_c = 0$: In this case the clutter samples are statistically independent; i.e., $\beta_n = 1$, $\forall n = 1, 2, \ldots, M\,(\mathbf{\Lambda_c} = \mathbf{I})$, and therefore equations (3.20) and (3.21) reduce, respectively, to

$$\Phi_{q|H_0}(s) = \frac{1}{(1 + s)^M} \quad (3.22)$$

and

$$\Phi_Q(s) = \frac{1}{(1 + s)^{MN}} \quad (3.23)$$

The poles of the mgf $\Phi_{q/H_0}(s)$ of q under hypothesis H_0 are a simple pole at $s = -1$ of multiplicity M lying in the left-hand s-plane. Thus, Substituting equations (3.22) and (8.23) into equation (3.12), we can write

$$P_{fa}\big|_{\rho_c=0} = -\frac{1}{(M-1)!}\frac{d^{M-1}}{ds^{M-1}}\left[\frac{1}{s\,(1-T_{CA}s)^{MN}}\right]_{s=-1} \tag{3.24}$$

Equation (3.24) can be expressed as [9]

$$P_{fa}\big|_{\rho_c=0} = \sum_{i=1}^{M}\binom{MN+i-2}{MN-1}\frac{T_{CA}^{i-1}}{(1+T_{CA})^{MN+i-1}} \tag{3.25}$$

We denoted T as T_{CA} since, in this case, the clutter plus noise samples are assumed IID. Equation (3.25) is the well known expression for the probability of false alarm of the ML detector processing M pulses.

Case 2. $\rho_c = 1$: In this case, the clutter samples are completely correlated; i.e., $\beta_1 = M$ and $\beta_n = 0$ for $n = 2,\ldots,M$ (Λ_c is singular), and therefore equations (3.20) and (3.21) reduce, respectively, to

$$\Phi_{q/H_0}(s) = \frac{1}{(1+Ms)} \tag{3.26}$$

and

$$\Phi_Q(s) = \frac{1}{(1+Ms)^N} \tag{3.27}$$

The poles of the mgf Φ_{q/H_0} of q under hypothesis H_0 are a simple pole at $s = -\frac{1}{M}$ lying in the left-hand s-plane. Hence, substituting equations (3.26) and (3.27) into equation (3.12), we can write

$$P_{fa}\big|_{\rho_c=1} = \frac{1}{(1+T)^N} \tag{3.28}$$

The result of equation (3.28) can also be obtained intuitively. Indeed, if $\rho_c = 1$, the clutter samples are expected to be the same (pair wise) from one pulse to another. Therefore, the first pulse contains all the average signal clutter power and the remaining pulses contain no signal clutter. In other words, we are in a single pulse situation in which equation (3.28) is readily obtained.

Case 3. $0\langle\rho_c\langle1$: In this case the clutter returns are partially correlated. The poles of the mgf $\Phi_{q|H_0}$ of q under hypothesis H_0 are at $s_n = -\frac{1}{\beta_n}$, n=1, 2,..., M. They are all distinct (Λ_c is a symmetric Toeplitz matrix) and all lie in the left-hand s-plane. Hence, substituting equations (3.20) and (3.21) into equation (3.12) and assuming that the correlation coefficient ρ_c is known, we can write, after some simple mathematical manipulations,

$$P_{fa} = \sum_{i=1}^{M}\left[\prod_{j=1,\,j\neq i}^{M}\left(1-\frac{\beta_j}{\beta_i}\right)^{-1}\prod_{n=1}^{M}\left(1+T\frac{\beta_n}{\beta_i}\right)^{-N}\right] \tag{3.29}$$

Equation (3.29) shows that the actual probability of false alarm of the ML-CFAR detector is a function of the number of reference cells N, the number of processed pulses M, the threshold multiplier T and the correlation coefficient ρ_c. That is, for every value of the threshold T, there is a unique value of ρ_c so that the actual probability of false alarm is equal to the prescribed probability of false alarm. Therefore, the search for the threshold multiplier T can be summarized by the following algorithm.

a) We assign a desired value of the P_{fa} and a given value of ρ_c.
b) We assume an initial value of T.
c)

     i)   If $\rho_c = 0$ then, we compute the P_{fa} using equation (3.25).
     ii)   If $\rho_c = 1$ then, we compute the P_{fa} using equation (3.28).
     iii)   If $0\langle\rho_c\langle1$ then, we compute the P_{fa} using equation (3.29).

d) If the value of the P_{fa} from step c) is equal to the desired value of the P_{fa} stop, otherwise go to step b).

The above procedure requires the knowledge of ρ_c which, in general, is not available a priori. The value of ρ_c used in step a) is the estimated value of the correlation coefficient between pulse-to-pulse neighboring clutter samples [5]. However, it is practically reasonable to assume a known correlation coefficient ρ_c since the two-pulse moving target indication (MTI) causes the clutter samples for a given range cell to be correlated from pulse-to-pulse and this correlation can be easily determined [7]. This is the case we want to confine our attention to.

3.4 Simulation Results

To evaluate the false alarm properties of the proposed model, we conducted computer simulations. All clutter samples were generated from the correlation models given by equation (3.16). We assume a reference window size of N=16 and design $P_{fa} = 10^{-4}$. First, the threshold multipliers T are computed using the algorithm listed in Section 3.3.2 to achieve the prescribed P_{fa} for values of ρ_c going from 0 to 1, and for values of the number of processed pulses $M = 2, 4$ and 8. Fig. 28.2, shows a set of curves of T against ρ_c. We observe that, for the prescribed P_{fa}, all curves increase monotonically with ρ_c to converge to the same value at $\rho_c = 1$. Therefore, for a given M, a variety of choices for T are then possible. The choices of the threshold multipliers that assume the actual clutter correlation provide a good regulation of the P_{fa} as shown in Fig. 28.3. However, the curves for $\rho_c = 1$ and $\rho_c = 0$, do not assume that the clutter correlation is changing, and therefore the prescribed P_{fa} is not achieved. Our ML-CFAR detector assumes knowledge of the value of ρ_c and then computes the corresponding T.

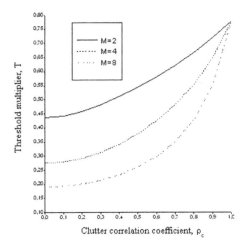

Fig. 3.2. Threshold multiplier against clutter correlation coefficient for $N = 16$ and $P_{fa} = 10^{-4}$.

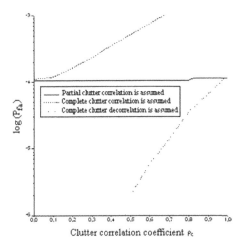

Fig. 3.3. Simulated probability of false alarm against clutter correlation coefficient for $N = 16$ and $P_{fa} = 10^{-4}$.

3.5 Conclusion

In this chapter, we have considered the problem of ML-CFAR regulation processing a pulse-to-pulse partially Rayleigh correlated clutter. Assuming that the clutter power is much higher than the noise power, we have derived a closed form for the probability of false alarm. To analyze the detection performance of the ML-CFAR detector, we have applied these results to show the target detectability as a function of the correlation between Gaussian target returns of all fluctuating models and multiple target situations [11].

References

1. Barnard T J, Weiner D D (1996) Non-Gaussian Clutter Modeling with Generalized Spherically Invariant Random Vectors
   IEEE Transactions On Signal Processing 44:2384–2390
2. Farina A, Gini F, Greco M, Verrazzani L (1997) High Resolution Sea Clutter Data: A Statistical Analysis of Recorded Live Data
   IEE Proceedings-F Vol. 144 3:121–130
3. Nohara T, Haykin S (1991) Canadian East Coast radar trials and the K-distribution
   IEE Proceedings-F Vol 138 2:80–88
4. Farina A, Russo A (1986) Radar Detection of Correlated Targets in Clutter
   IEEE Transactions on Aerospace and Electrical Systems Vol 22 5:513–532
5. Himonas S D, Barkat M (1990) Adaptive CFAR Detection in Partially Correlated Clutter
   IEE Proceedings-F Vol 137 5:387–394
6. Al Hussaini E K, El Mashade M B (1996) Performance of Cell-averaging and Order-statistic CFAR Detectors Processing Correlated Sweeps for Multiple Interfering Targets
   Signal Processing 49:111–118
7. El Mashade M B(2000) Detection Analysis of CA Family of Adaptive Radar Schemes Processing M-correlated Sweeps in Homogeneous and Multiple-target Environments
   Signal Processing 80:787–801
8. Ritcey J A (1985) Calculating Radar Detection Probabilities by Contour Integration. California
   PhD dissertation, University of California, San Diego
9. Hou Xiu-Ying et al. (1987) Direct Evaluation of Radar Detection Probabilities
   IEEE Transactions on Aerospace and Electrical Systems Vol. AES-23 4:418–423
10. Barkat M (2005) Signal Detection and Estimation. In: Artech House (Second Edition) Boston
11. Laroussi T, Barkat M (2007) Adaptive ML-CFAR Detection for Correlated Chi-square Targets of all Fluctuation Models in Correlated Clutter and Multiple Target Situations
    ISSPA Sharjah United Arab Emirats

4

On-orbit Spatial Resolution Estimation of CBERS-2 Imaging System Using Ideal Edge Target

Kamel Bensebaa[1], Gerald J. F. Banon[1], Leila M. G. Fonseca[1], and Guaraci J. Erthal[1]

Image Processing Division - National Institute for Space Research (INPE) Av. dos Astronautas, 1758, 12227-010 São José dos Campos, Brazil phone: 55 12 39 45 65 22, fax: 55 12 39 45 64 68 {camel,banon,leila,gaia}@dpi.inpe.br

Summary. The China-Brazil Earth Resources Satellite (CBERS-2) has been developed by China and Brazil and was launched on October 2003. This satellite carries three sensors: WFI, CCD and IRMSS. Due to limitations of the CCD sensor components, the images acquired by the imaging system undergo a blurring. Under the hypothesis that the Point Spread Function (PSF) of the imaging system is Gaussian, the blurring effect of an ideal step edge can be represented as an error function (erf). As the PSF is assumed separable, its identification reduces to the estimation of two standard deviations or equivalently to two EIFOVs (Effective Instantaneous Field of View), one for the along-track direction and another for the across-track. This work describes an approach for the on-orbit CBERS-2 CCD spatial resolution estimation using a set of subimages of natural edges and allows an objective assessment of the imaging system performance.

Key words: CCD camera, spatial resolution, estimation, modelling, target, erf, edge spread function, point spread function, EIFOV.

4.1 Introduction

A cooperative remote-sensing program between Brazil and China initiated in 1988, has resulted in the development and the building up a set of remote sensing satellites called CBERS (China-Brazil Earth Resources Satellite). These satellites allow monitoring their huge territories, mainly: environmental change, ground survey, natural disaster, agriculture and deforestation.

The first stage of this program was accomplished by the launching of CBERS-1 on October 1999, which operated until August 2003. The second stage of the program consisted of the launching on October 2003 of CBERS-2, technically similar to its predecessor, with only minor changes to ensure its reliability.

Imagery from orbiting sensors has provided much information about the Earth's surface and the effects of human activities upon it. For this information be useful, it is critical to access the imagery system performance. One performance measure is related to the blurring effect due to the instrumental optics (diffraction, aberrations, focusing error) and the movement of the satellite during the imaging process.

Hence, the images may have a blurred appearance that is likely to compromise their visual quality and analysis. In this sense, the performance evaluation of imaging system in term of spatial resolution estimation is an important issue.

In general, the blurring effect is related to the Point Spread Function (PSF) in the spatial domain and to the Modulation Transfer Function (MTF) in the frequency domain [6]. For translation invariant linear system, the PSF characterizes the imaging system. Under the assumption that the PSF is Gaussian and separable, its identification reduces to the estimation of two parameters called EIFOV (Effective Instantaneous Field of View), one for the along-track direction and another for the across-track which are equal to 2.66 times the standard deviation [2, 16]. This estimation allows spatial resolution estimation, and consequently an objective assessment of the imaging system performance. The EIFOV enables a comparison between different sensors.

Among other approaches, the spatial resolution of an imaging system may be obtained from the blurring effect of an ideal step edge. In natural scenes, edges are not always ideal step edges. For that reason, only the "better" edges are selected. When the imaging system is excited by an ideal step edge, the transition from bright to dark defines the edge sharpness and it is used to estimate the spatial resolution.

This transition is represented by the so-called Edge Spread Function (ESF) [12]. Despite the fact that this function is 2D, it can be characterized through a 1D function along the normal of the edge in the case of translation invariant systems. Furthermore, when the system is linear and the PSF is Gaussian, this function is an error function (erf) which is the convolution product of the ideal step edge with a Gaussian function.

The two EIFOV characteristic parameters of the spatial resolution can be theoretically obtained from two ESFs in different directions. In practice more than two ESFs are convenient to get a more precise estimation.

The objective of this work is to use an approach for an on-orbit assessment performance of the CCD camera of CBERS-2 satellite which doesn't depend on any target size measurement. The approach consists of estimating the along-track and across-track EIFOVs using a set of subimages of natural edges extracted from a scene image of Sorriso town in Mato Grosso state (Brazil). Each selected subimage corresponds to a different edge direction.

A standard deviation associated with each subimage was estimated and the spatial resolution estimation of the along-track and across-track was obtained through an ellipse fitting.

The rest of the Chapter is organized as follows: Section 4.2 gives a brief overview of the CBERS-2 satellite. Section 4.3 summarizes four different PSF estimation approaches. Section 4.4 introduce respectively the target scene of natural edges between different crops or between crops and nude soils, and the data preparation. Section 4.5 describes the algorithm used in this work for edge detection and edge cross-section extraction, and it presents the edge model. Section 4.6 introduces the details of the ellipse fitting technique. Finally, the Sect. 4.7 discusses the results and gives the conclusion of this work.

4.2 CBERS-2 Overview

CBERS-2 satellite carries on-board a multisensor payload with different spatial resolutions called: WFI (Wide Field Imager), IRMSS (Infrared MSS) and CCD (Charge Coupled Device) camera. In addition, the satellite carries a Space Environment Monitor for detecting high-energy radiation.

The high-resolution CCD Camera device which is the main study of this work, provides images of 4 spectral bands from visible light to near infrared (B1: 0.45–0.52 μm; B2: 0.52–0.59 μm; B3: 0.63–0.69 μm; B4: 0.77–0.89 μm) and one panchromatic band (B5: 0.51–0.73 μm). It acquires the earth ground scenes by pushbroom scanning, from a 778 km sun-synchronous orbit and provides images of 113 km wide strips with sampling rate of 20 meters at nadir for the multispectral bands.

4.3 PSF Estimation Approaches

Basically, there are four approaches to determine a PSF or a MTF of an imaging system. They are based on experimental methods or in theoretical modelling of the physical processes under study.

The first approach uses the image system specifications to model its spatial response. Fonseca and Mascarenhas [8] and Markham [15] have used this approach to determine the spatial response of the TM sensor (Landsat satellite).

The second approach uses targets with well-defined shape and size as airport runway, bridges, artificial targets, etc. For these targets, Storey [17] has provided a methodology for on-orbit spatial resolution estimation of Landsat-7 ETM+ sensor by using a Causeway Bridge image (Louisiana – USA). Choi and Helder [7] have used airport runway and a tarp placed on the ground for on-orbit MTF measurement of IKONOS satellite sensor. Bensebaa et al. [4] used an image of an artificial black squared target placed in the Gobi desert (China). The CCD spatial response is modelled as 2D Gaussian function which is characterized by two parameters: one in along-track direction and another in across-track. The EIFOV values are then derived from these parameters. Bensebaa et al. [3] also used natural targets such as the Rio-Niter Bridge over Guanabara Bay (Brazil) and Causeway Bridge to estimate the spatial resolution in the along-track and across-track directions, respectively.

As opposed to the second approach, in the third approach the target size doesn't need to be known since the target consists of an ideal step edge. This approach was already successfully experimented by Luxen and Förstner [14].

The fourth approach consists of adjusting a simulated low resolution image to an image acquired by the sensor under study. According to Storey [17], this method works satisfactorily if the two sets of imagery are acquired at or near the same time or, at least, under similar conditions to avoid the problems associated with temporal variations. This kind of experiment was used to determine the spatial resolution of the CBERS-CCD cameras using a higher spatial resolution image acquired by the SPOT-4 satellite and an image of the same scene acquired by CBERS satellite [5].

The algorithm proposed in this work belongs to the third approach but differently from the work of Luxen and Förstner [14], the point spread function model used here, may be assumed separable since the selected images are raw data whose rows correspond to the CCD chips and columns correspond to the detectors movements both being in orthogonal directions.

4.4 Target Image

The initial task is the selection of natural edges between different crops or between crops and nude soils. In this sense, the scene of Sorriso town located 270 miles north of Cuiabá, the state capital of Mato Grosso (Brazil) is perfect. Sorriso County now plants above 700,000 acres of soybeans annually. In addition, this region is also a producer of corn and cotton. Figure 4.1 shows original image of Sorriso region.

This work used band B2 of the scene of Sorriso with orbit 116 and point 114, acquired on July 15, 2006 by CCD camera on-board CBERS-2. The original cloud free image was a good candidate for the extraction of several subimages of different edge directions. Twelve of these subimages were selected for this work.

4.5 Edge Processing

In this section, we describe the algorithm for edge cross-section extraction as well as the edge model. The illustrations are done, using a subimage with edge direction of 143° .

4.5.1 Edge detection and edge cross-section extraction

The first step of the algorithm is the edge detection. For this step, the Sobel operator [10] was used with a thresholding operation that results in a binary

Fig. 4.1. Original image of Sorriso region (band 2)

image. This operator was chosen because it's less sensitive to isolated high intensity point. It is a "small bar" detector, rather than a point detector. Figure 4.2(a) shows the original subimage and Fig. 4.2(b) shows the result of Sobel edge operation after thresholding.

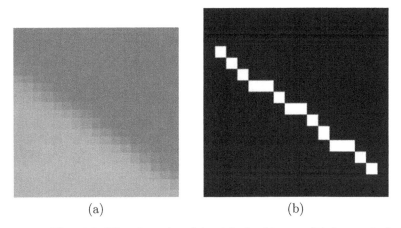

(a)   (b)

Fig. 4.2. Edge detection: (a) original subimage; (b) detected edge

Once the edge detection operation is performed, its gravity center G_c is computed using the following expression:

$$G_c = \frac{\sum_{(u,v)} (u,v).I_E(u,v)}{\sum_{(u,v)} I_E(u,v)} \tag{4.1}$$

where (u,v) represents the position of each pixel in the binary edge image I_E and $I_E(u,v)$ represents its radiometry.

A 7×7 subimage was extracted in such a way that its center coincides with the previously computed gravity center. This operation allows the centering of the detected edge in the subimage. Figure 4.3(a) shows the centralized subimage and the Fig. 4.3(b) shows the centralized detected edge.

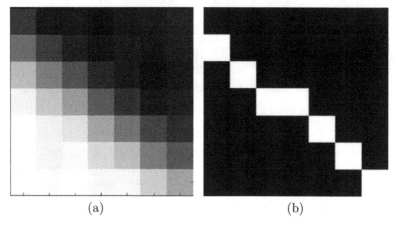

(a)   (b)

Fig. 4.3. Edge centralization: (**a**) centralized subimage; (**b**) centralized detected edge

The parameters a, b and c of the edge fitting straight line equation:

$$au + bv + c = 0 \tag{4.2}$$

were estimated by solving the homogeneous linear equation system given by expression:

$$\mathbf{Ax} = \mathbf{0} \tag{4.3}$$

where $\mathbf{x}$ is the vector of 3 unknowns $[a\ b\ c]^t$, $\mathbf{0}$ is the null vector $[0\ 0\ 0]^t$ and $\mathbf{A}$ is the matrix $M \times 3$ with $M > 3$ given by:

$$\mathbf{A} = \begin{bmatrix} u_1 & v_1 & 1 \\ \vdots & \vdots & \vdots \\ u_M & v_M & 1 \end{bmatrix}, \tag{4.4}$$

where the $(u_i, v_i)_{i=1,..,M}$ are the coordinates of the edge pixels.

The best way to solve (4.3) is to perform singular value decomposition on the matrix $\mathbf{A}$ [13].

The next step consists of extracting the edge cross-section along the edge normal. This edge cross-section is the function E_R that maps the real numbers:

$$\rho(u,v) = \frac{(au + bv + c)}{\sqrt{a^2 + b^2}} \qquad (4.5)$$

to the radiometries I_R, when (u,v) runs over the domain of the subimage I_R (7 × 7 set of points). Actually $\rho(u,v)$ represents the distance of the pixels position (u,v) to the edge straight line given by (4.2) [1].

Figure 4.4 depicts the edge cross-section extracted from a given subimage. The domain unit is meter and the range unit is radiometry digital number.

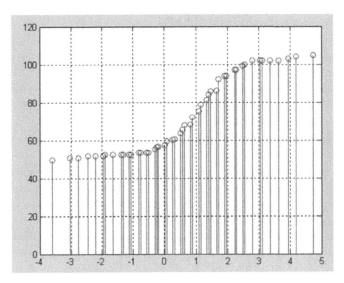

Fig. 4.4. An edge cross-section

4.5.2 Edge Model

Let M_1 and M_2 represent, respectively the least and the greatest values of the edge cross-section, then the edge model is given by:

$$E_M(\rho) = \left(1 - \frac{1}{2} * erfc\left(\frac{\rho - \mu}{\sigma\sqrt{2}}\right)\right) * (M_2 - M_1) + M_1 \qquad (4.6)$$

where *erfc* represents the complementary error function given by:

$$erfc(x) = \frac{2}{\sqrt{\pi}} \int_x^\infty e^{-t^2} dt \qquad (4.7)$$

In (4.6), the parameters μ and σ represent respectively the mean and the standard deviation of the underlying Gaussian function.

Let $\mathrm{RMS}(E_R, E_M)$ be the root mean square of the difference between E_R and E_M:

$$\mathrm{RMS}(E_R, E_M) = \left(\frac{1}{\#\mathrm{Domain}(E_R)} \sum (E_R - E_M)^2 \right)^{1/2} \qquad (4.8)$$

The estimation procedure has two-steps. At the first step for a given default value σ , we look for μ which minimizes $\mathrm{RMS}(E_R, E_M)$. At the second step, we use the previous optimal value μ and look for σ which minimizes $\mathrm{RMS}(E_R, E_M)$. The optimal values have been obtained by nonlinear programming [11].

Figure 4.5 depicts the result of the fitting of the edge model over the edge cross-section for an edge direction of 143° .

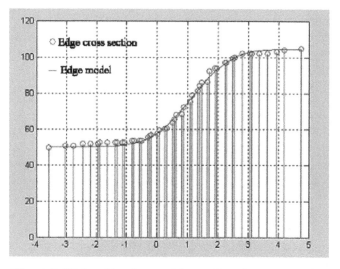

Fig. 4.5. Fitting the edge model over the edge cross-section

4.6 Ellipse Fitting and EIFOV estimation

The last step is the ellipse fitting. More specifically, because of the separability assumption of the PSF, the problem is the fitting of the ellipse given by (4.9):

$$\frac{x^2}{\sigma_x^2} + \frac{y^2}{\sigma_y^2} = 1 \qquad (4.9)$$

over the set of points (x_i, y_i) $i = 1, ..., 12$ where

$$x_i = \sigma_i \frac{|b_i|}{\sqrt{a_i^2 + b_i^2}} \tag{4.10}$$

$$y_i = \sigma_i \frac{|a_i|}{\sqrt{a_i^2 + b_i^2}} \tag{4.11}$$

and where σ_i is the optimal standard deviation for edge i with parameters a_i and b_i. Each point (x_i, y_i) characterizes the blurring effect in the direction of the normal to the edge i, its distance to the origin is σ_i.

The above expression are for an edge direction $arctg(-a_i/b_i)$ comprised between 0 and $\pi/2$. For the other quadrants some appropriate signal must be added in these expressions.

The parameters σ_x and σ_y are estimated by solving the homogeneous linear equation system given by the following expression:

$$\begin{bmatrix} x_1^2 & y_1^2 & -1 \\ \vdots & \vdots & \vdots \\ x_{12}^2 & y_{12}^2 & -1 \end{bmatrix} \cdot \begin{bmatrix} \alpha \\ \beta \\ \gamma \end{bmatrix} = 0 \tag{4.12}$$

The estimated values of standard deviations σ_x and σ_y in along-track and across-track directions are respectively,

$$\sigma_x = \sqrt{\frac{\gamma}{\alpha}} \quad \text{and} \quad \sigma_y = \sqrt{\frac{\gamma}{\beta}}. \tag{4.13}$$

Finally, the optimal EIFOV values for both directions are related to the standard deviation σ by the expression:

$$EIFOV = 2.66.\sigma \tag{4.14}$$

Results of the optimal values of the standard deviation and the EIFOVs are presented in Table 4.1.

Table 4.1. Results of optimal EIFOVs

Direction	Standard Deviation (m)	EIFOV (m)
Along-track	19.20	51
Across-track	25.26	67

Figure 4.6(a) shows the subimages and its corresponding EIFOV while Fig. 4.6(b) shows the best fitting of the ellipse.

4.7 Conclusion

In this Chapter, an approach for CBERS-2 CCD on-orbit spatial resolution estimation has been introduced using subimages of natural edges extracted

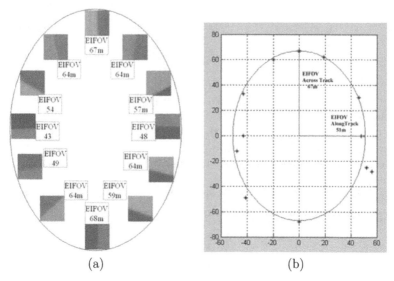

(a)   (b)

Fig. 4.6. EIFOV estimation: (**a**) selected subimages; (**b**) ellipse fitting

from the original image of a scene of Sorriso town in Mato Grosso state in Brazil.

The results show that the CBERS-2 CCD across-track resolution is worse than the along-track one and confirm the results obtained in previous works. This degradation could be explained by the presence of mirror vibration when both sensors IRMSS and CCD work simultaneously.

Besides this hypothesis, the observed degradation could be the result of an electronic coupling between adjacent detectors. In addition, we have noticed a little degradation of the EIFOV in along-track direction in comparison with the previous results [3, 4, 5], even though this method of using edges as targets leads to conservative evaluations because of the difficulty of finding ideal edges.

This EIFOV estimation result is valuable in future work on CBERS-2 image restoration.

References

1. Banon GJF (1989) Bases da computação gráfica.Editora Campus Ltda, Rio de Janeiro
2. Banon GJF, Santos AC (1993) Digital filter design for sensor simulation: application to the Brazilian Remote Sensing Satellite. São José dos campos: INPE. 62 p. (INPE-5523- RPQ/665). Available from: http://bibdigital.sid.inpe.br/rep/dpi.inpe.br/banon/1995/12.14.18.12 Access date: 29 oct. 2003.
3. Bensebaa K, Banon GJF, Fonseca LMG (2004) On-orbit Spatial Resolution Estimation of CBERS-1 CCD Imaging System from bridge images. In: Congress

International Society for Photogrammetry and Remote Sensing, 20., 12-23 July 2004, Istanbul, Turkey. Proceedings... Istambul: International Archives of Photogrammetry, Remote Sensing and Spatial Information Sciences, v. 35, part B1, p. 36-41.

4. Bensebaa K, Banon GJF, Fonseca LMG (2004) On-orbit spatial resolution estimation of CBERS-1 CCD camera. In: International Conference on Image and Graphics (ICIG 2004), 3., 18-20 dec. 2004, Hong Kong, China. Proceedings... Los Alamitos, CA: IEEE Computer Society Press. v. 2 p. 576-579.

5. Bensebaa K, Banon GJF, Fonseca LMG (2005) On-orbit spatial resolution estimation of CBERS-1 CCD imaging system using higher resolution images. In: Simpósio Brasileiro de Sensoriamento Remoto (SBSR), 12., 2005, Goiânia. Anais... São José dos Campos: INPE, 2005. Artigos, p. 827-834. CD-ROM. ISBN 85-17-00018-8. Available from: http://bibdigital.sid.inpe.br/rep/ltid.inpe.br/sbsr/2004/11.19.19.28 Access date: 12 may 2005.

6. Bretschneider T (2002) Blur identification in satellite imagery using image doublets. In: Asian Conference on Remote Sensing, 23., Kathmandu, Nepal. Proceedings... Kathmandu: GIS Development, 2002, CD-ROM. Available from: http://www.gisdevelopment.net/aars/acrs/2002/adp/119.pdf Access date: 11 apr. 2004.

7. Choi T, Helder D (2003) On-Orbit Modulation Transfer Function (MTF) Measurement. Remote Sensing of Environment 88:42–45

8. Fonseca LMG, Mascarenhas NDD (1987) Determinação da função de transferência do sensor TM do satélite Landsat-5. In: Congresso Nacional de Matemtica Aplicada e Computacional, 10., 21-26 set. 1987, Gramado, BR. Resumo dos Trabalhos... São José dos Campos: INPE, 1987. p. 297-302. Publicado como: INPE-4213-PRE/1094

9. Gander W, Golub GH, Strebel R (1994) Least-squares fitting of circles and ellipses, Bit, 34:558–578

10. Gonzalez RC, Woods RE (2002) Digital Image Processing 2nd Ed. Prentice Hall, New Jersey

11. Himmelblau DM (1972) Applied nonlinear programming. McGraw-Hill, New York

12. Kohm K (2004) Modulation Transfer Function Measurement Method and Results for The Orbview-3 High Resolution Imaging Satellite. In: Congress International Society for Photogrammetry and Remote Sensing, 20., 12-23 July 2004, Istanbul, Turkey. **Proceedings**... Istambul: International Archives of Photogrammetry, Remote Sensing and Spatial Information Sciences, v. 35, part B1, p. 7–12

13. Kreyszig E (1993) Advanced engineering mathematics. John Wiley and Sons, New York

14. Luxen M, Förstner W (2002) Characterizing image quality: Blind estimation of the point spread function from a single image. In: ISPRS Commission III Symposium (Photogrammetric Computer Vision), 9-13 Setember 2002, Graz, Austria. **Proceedings**... Graz: PCV02, 2002, v. XXXIV, part 3A, p. 205–210

15. Markham BL (1985) The Landsat sensor's spatial response. IEEE Transactions on Geoscience and Remote Sensing, 23:864–875

16. Slater PN (1980) Remote sensing optics and optical system. Addison-Wesley, London

17. Storey JC (2001) Landsat 7 on-orbit modulation transfer function estima-
    tion. In: Sensors, Systems, and Next Generation Satellites 5., 17-20 sep. 2001,
    Toulouse, France. **Proceedings**...Bellingham, WA, USA: SPIE, 2001, p. 50–
    61

Distributed Pre-Processed CA-CFAR Detection Structure For Non Gaussian Clutter Reduction

Zoubeida Messali[1] and Faouzi Soltani[2]

[1] Laboratoire Signaux et Systèmes de Communication, Département d'Electronique, Université de Constantine 25010, Algeria.messalizoubeida@yahoo.fr

[2] Laboratoire Signaux et Systèmes de Communication, Département d'Electronique, Université de Constantine 25010, Algeria.f.soltani@caramail.com

Summary. This chapter deals with the cell averaging constant false alarm rate (CA-CFAR) distributed RADAR detection of targets embedded in a non Gaussian clutter. We develop a novel pre-processing algorithm to reduce spiky clutter effects, notably a non linear compression procedure. This technique is similar to that used in non uniform quantization where a different law is used. Two approaches to combine data from the pre-processed CA-CFAR detectors are proposed The performance characteristics of the proposed CA-CFAR distributed systems are presented for different values of the compression parameter. We demonstrate, via simulation results, that the pre-processed procedure combined with distributed systems when the clutter is modeled as alpha-stable distribution, are computationally efficient and can significantly enhance detection performance.

Key words: Distributed radar detection, Non gaussian clutter, Non linear Compression

5.1 Introduction

In recent years, multisensor data fusion has received significant attention for military and non military applications [1]. The RADAR network is a kind of multisensor data fusion system. An important point in RADAR detection is to maintain a constant false alarm rate when the background noise fluctuates. Hence, CFAR detectors have been designed to set the threshold adaptively according to local information on the background noise. The CA approach is such an adaptive procedure. The design of a distributed CFAR detection system is strongly affected by the clutter model assumed. A number of studies suggest that clutter modeling is more accurately achieved by heavy-tailed distributions than Rayleigh and Weibull distributions, such as active sonar returns, sea clutter measurements, and monostatic clutter from the US Air Force Mountaintop Database [2]. Indeed, alpha-stable processes have to be effective in modeling many real-life engineering problems [3,4] such as outliers and impulsive signals [2]. We will use a non linear compression method for spiky clutter reduction. This new algorithm consists on compressing the output square law detector noisy signal with respect to a non linear law in order to reduce the effect of impulsive noise level. We present two configurations to combine data from multiple CA CFAR detectors based on a non linear compression procedure for spiky clutter reduction. The organization of the chapter is as follows: Section 5.2, presents our method of non gaussian clutter reduction for single CA-CFAR detector. We derive the false alarm probability P_{FA} of CA-CFAR detector based on a non linear compression procedure for alpha-stable measurements. In section 5.3, we present a distributed system based on the proposed method to achieve even better performance. Two approaches to combine data from the pre-processed CA-CFAR detectors are discussed. Finally, the results and conclusions are provided in sections 5.4 and 5.5 respectively.

5.2 The Pre-Processed CA-CFAR Based On A Non Linear Compression (PCA-CFAR) For Non Gaussian Clutter Reduction

We suppose that the clutter is modeled as alpha-stable distributed data. We propose to pass the noisy signal through a non linear device that compresses the large amplitude (i.e., reduces the dynamic range of the noisy signal) before further analysis as proposed in [5]. Fig 5.1, illustrates the pre-processed CA-CFAR detector (PCA-CFAR) block diagram. X and $\tilde{X}$ are the square law detector output and the compressed signal respectively. The output of the non linear device is expressed as:

$$\tilde{X} = g\ [X] = |\ X\ |^{\ \beta}.sign\ [X] \tag{5.1}$$

where $0 < \beta \ \leq 1$ is a real coefficient that controls the amount of compression applied to the input signal of the non linear compressor. This technique is similar to that used in non uniform quantization where a different non linear law is used [6].

In the following we derive the expression of the pdf of the statistic at the output of the non linear compressor. The statistic $y = \tilde{x}$ is the output of the compressor, then

$$y = \tilde{x} = g\,(x) = |x|^{\ \beta}.sign\,(x) \tag{5.2}$$

where x is the input of the compressor and is Pearson distributed. The pdf, $p_X(x)$ is given by

$$p_{X_i}(x) = \begin{cases} \frac{\gamma}{\sqrt{2\pi}} \frac{1}{x^{3/2}} e^{-\gamma^2/2x}, & x \geq 0 \\ 0, & otherwise \end{cases} \qquad (5.3)$$

where γ is the dispersion of the distribution.
$p_y(y)$ is given by [7]

$$p_Y(y) = \frac{f_X(x)}{|g'(x)|} \qquad (5.4)$$

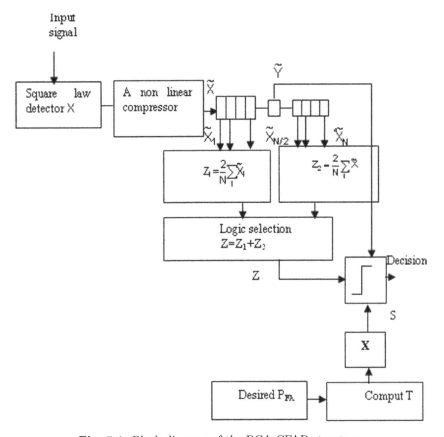

Fig. 5.1. Block diagram of the PCA-CFAR structure.

substituting x by $(y^{1/\beta})$, (5.4) results in

$$p_Y(y) = \frac{\gamma}{\sqrt{2\pi}\beta} \frac{1}{y^{\frac{2\beta+1}{2\beta}}} e^{-\frac{\gamma^2}{2y^{1/\beta}}} \qquad (5.5)$$

The probability of false alarm P_{FA}^{PCA} can be expressed as

$$P_{FA}^{PCA} = \Pr\left\{\tilde{Y}_0 \geq TZ\right\} = \int_0^\infty \Pr\left\{\tilde{Y}_0 \geq Tz\right\} p_z(z)dz \qquad (5.6)$$

where $\tilde{Y}_0$ is the compressed noise random variable, interpreted as a target echo during the thresholding decision. The statistic Z is the estimate of the average clutter level after compression and equal to

$$Z = \left(\frac{1}{N}\right) \sum_{i=1}^{N} \tilde{X}_i \qquad (5.7)$$

T is a scaling factor used to achieve a certain P_{FA}.

We should note here that the scaling factor T is found by simulation based on Monte Carlo counting procedure for a fixed P_{FA}^{PCA} ($P_{FA}^{PCA} = 10^{-4}$) with ten thousands iterations.

P_{FA}^{PCA} is controlled by the scaling factor T and it does not depend on the dispersion parameter γ of the Pearson-distributed parent population. Hence, the proposed PCA is also a CFAR method for compressed Pearson background. For $\beta=1$ (i.e., no compression), P_{FA}^{PCA} is equal to P_{FA} of the conventional CA-CFAR detector.

The corresponding probability of detection (P_D) for the case of a Rayleigh fluctuating target with parameter σ_s^2 in a heavy-tailed background noise can be expressed as

$$P_D^{PCA} = \Pr\left\{\tilde{Y}_1 \geq TZ\right\} = \int_0^\infty \Pr\left\{\tilde{Y}_1 \geq Tz\right\} p_z(z)dz \qquad (5.8)$$

where

$$\tilde{Y}_1 = |Y_1|^\beta \, sign(Y_1) \qquad (5.9)$$

An exact analytical evaluation of the expression (8) is not possible. In fact, to specify Y_1 under H_1 would require specifying the in-phase and quadrature components of both the clutter and the useful signal, whereas only their amplitudes pdfs are given. Therefore, we have to resort to computer simulation to evaluate P_D^{PCA}. Hence, the test-cell measurement is considered as a compressed scalar product of the two vectors: the clutter and the useful signal respectively. So that

$$Y_1 = s + c + \sqrt{s.c} \cdot cos\,(\theta) \qquad (5.10)$$

where θ is the angle between the vectors s and c and is uniformly distributed in $[0,2\pi]$. s and c are the input signal and clutter components, respectively. Note that P_D is a function of the clutter dispersion γ, the power parameter of the Rayleigh fluctuation target σ_s and the compressor parameter β.

In the following we propose two configurations to combine data from the pre-processed CA-CFAR detectors.

5.3 Distributed PCA-CFAR System For Non-Gaussian Clutter Reduction

Two approaches to combine data from the pre-processed CA-CFAR detectors are discussed in this section. In the fusion configuration shown in Fig. 5.2, we consider a distributed detection system constituted of n independent CA-CFAR sensors based on the non linear compression method. The clutter is modeled as alpha-stable distributed data. Each detector i, i=1,2,...n, based on a compressed observation vector $\tilde{X}_i$ makes a decision cu_i. Each decision cu_i may take the value 0 or 1 depending on whether the detector i decides H_0 or H_1. The global decision cu_0 is made based on the received decision vector containing the individual decisions i.e., $\mathbf{CA}$=(cu_1,cu_2,............cu_n) according to "OR" logic. The Neyman-Pearson (N-P) formulation of this distributed detection problem is considered.

In the fusion configuration, shown in Fig. 5.3, we compute the mean vector of the compressed observation vectors of all detectors. The resulting computed vector is then introduced into a CA-CFAR block. The global decision cu_0 is made according to CA-CFAR procedure and N-P formulation. The hope being is to enhance detection performances of the fusion configuration shown in Fig. 5.2. We assume that the detectors of the distributed systems have the same characteristics, i.e., equal P_{FA}^{PCA} and equal number of reference cells N_i. It is worth noting that almost no gain is achieved by adopting larger reference windows. We note also, that the combination and increasing the number of pre-processed sensors are more effective than enlarging the reference window, as far as the probability of detection is concerned. Hence, distributed pre-processed detectors combined with two fusion configurations and operating in alpha-stable background, behave considerably better than a single pre-processed sensor.

5.4 Results and Discussions

This section is devoted to the performance assessment of the proposed CA-CFAR detector. We consider the case of a Rayleigh fluctuating target embedded in Pearson distributed environment. We obtain through simulation results, based on Monte Carlo counting procedure, P_D versus the generalized signal-to-noise ratio (GSNR) for the case of Pearson clutter for a fixed P_{FA}. The GSNR is defined in [8] as:

$$GSNR = 20 \log \frac{\sigma_s}{\gamma} \tag{5.11}$$

where σ_s is the parameter of the Rayleigh fluctuating target.

We should note here that although the moment EX^2 of a second-order process has been widely accepted as standard measure of signal strength and associated with the physical concept of power and energy, it cannot be used with alpha-stable distributions because it is infinite. Hence, the GSNR expression in (5.14) should not be interpreted as a signal to noise power ratio. The results obtained are shown in Figs.5.4-6. Fig 5.4 illustrates the effect of the compression parameter β on P_D. The latter controls the amount of compression. As we can see the pre-processed PCA-CFAR achieves better performance than the CA-CFAR without pre-processing especially for low GSNR and for smaller values of β ($\beta=0.1$). For $\beta=1$ (no compression), the proposed PCA-CFAR gives the same results as the CA-CFAR. The scaling factor used to achieve a desired P_{FA} ($P_{FA}^{PCA}=10^{-4}$) for the pre-processed detector has been computed by simulation techniques for different values of β. In Figs. 5.5 and 5.6, we show the detection probability of both the two fusion configurations based on the non linear compression method for spiky clutter reduction. As we can see, the combination of the distributed system and the compression reduction procedure with two configurations is more effective than the single PCA-CFAR detector.

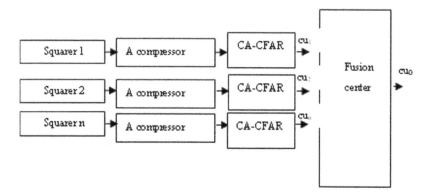

Fig. 5.2. First configuration of distributed PCA-CFAR detectors.

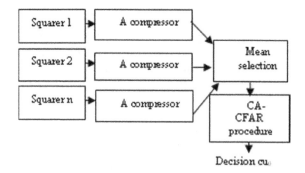

Fig. 5.3. Second configuration of distributed PCA-CFAR detectors.

5.5 Concluding Remarks

This chapter developed multisensor techniques based on a non linear compression procedure for non Gaussian clutter reduction . The performance of the proposed CA-CFAR detector operating in alpha-stable environment has been analysed. A non linear compression filter is introduced to reduce clutter spikiness. The key of this pre-processed method is the appropriate choice of β which controls the amount of compression. The comparisons of the proposed CFAR procedure with the conventional CA-CFAR detection have clearly demonstrated the superiority of the

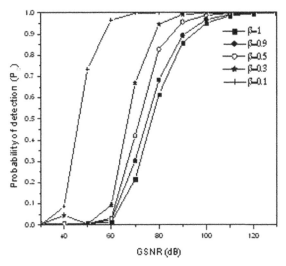

Fig. 5.4. P_D of single PCA-CFAR processor in homogeneous Pearson background as a function of GSNR=20log(σ_s/γ) and different values of β. N_{PCA}=32, P_{FA}^{PCA}= 10^{-4}.

pre-processed detector over the conventional CA processing especially for β=0.1. Therefore, the proposed pre-processed method is computationally efficient and can significantly enhance detection and reduce the noise effects. It has been shown, from simulation results, that the compression clutter reduction method does not have a significant effect on the detection probability for large values of β. The results obtained showed that the combination of the compression reduction procedure and multisensor parallel topology when the clutter is modeled as alpha-stable distributed data, has considerably increased the detection probability especially for small values of GSNR and for smaller values of β (β=0.1 and β=0.3).

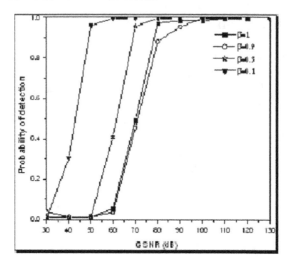

Fig. 5.5. P_D of distributed PCA-CFAR detectors adopting The 1st fusion configuration, as a function of GSNR=20log(σ_s/γ) and different.

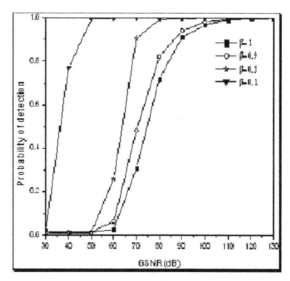

Fig. 5.6. P_D of distributed PCA-CFAR detectors adopting The 2nd fusion configuration, as a function of GSNR=20log(σ_s/γ) and different values of β. N_{PCA} =32, $P_{FA}^{PCA} = 10^{-4}$.

References

1. Waltz E, Lina J (1990) Multisensor Data Fusion. Chapter 1-12: Artech House, Boston
2. Nikias C L, Shao M (1995) Signal Processing with Alpha-Stable Distributions and Applications John Wiley
3. Pierce R D, (1997) Application of the Positive Alpha-Stable Distribution. IEEE Signal Processing Workshop on Higher-Order Statistics, Banff, Alberta, Canada:420-424
4. Tsakalides P, Nikias C L (1999) Robust Space-Time Adaptive Processing (STAP) in Non-Gaussian Clutter Environments. IEE Proceedings, Radar, Sonar and Navigation 14:84-94
5. Sahmoudi M, Abed-Meraim K, Barkat B (2004) IF Estimation of Multicomponent Chirp Signal in Impulsive a-Stable Noise Environment Using Parametric and Non-Parametric Approches. Proceeding of EUSIPCO'2004, Austria
6. Jayant N, Noll P (1984) Digital Coding Of Waveforms: Principles and Applications To Speech and Video: Prentice-Hall
7. Papoulis A (1991) Probability Random Variables and Stochastic Processes: McGraw-Hill Inc, 3rd Edn
8. Tsakalides P, Trinci P, Nikias C L (2000) Performance Assessment of CFAR Processors In Pearson-Distributed Clutter. IEEE Transactions on Aerospace and Electronic Systems, AES-36:1377-1386

6

Multispectral Satellite Images Processing through Dimensionality Reduction

Ludovic Journaux, Irène Foucherot and Pierre Gouton

LE2I, UMR CNRS 5158 Universitde Bourgogne, Aile des Sciences de l'Ingénieur, BP 47870 21078 Dijon Cedex, FRANCE
{ljourn, iFoucherot, pgouton}@u-bourgogne.fr

6.1 Introduction

Current satellite sensors provide multispectral images that are of great potential to discriminate between different types of landscape. Multivariate datasets are however difficult to handle: the information is often redundant, as spectral bands are highly correlated with one another. Satellite images are also typically large, and the computational cost of elaborate data processing tasks may be prohibitive. One possible solution is to use Dimensionality Reduction (DR) techniques. Their goal is to find a subspace of lower dimensionality than the original data space, on which to project the dataset, while retaining its features. They are useful to reduce the computation time of subsequent data analysis . They could also, by studying the shape of the manifold, produce insight into the process that created the data. Last but not least, they also help visualizing complex multidimensional data.

Numerous studies have aimed at comparing DR algorithms, usually using synthetic data. In [1], we took the operational viewpoint, by comparing 5 DR algorithms with real-life data. The methods compared in [1] were Principal Components Analysis, Curvilinear Components Analysis, Curvilinear Distance Analysis, ISOMAP (Isometric Feature Mapping) and Locally Linear Embedding. We applied these dimension reduction methods on multispectral Landsat 7 images and compared the results of a K-means algorithm applied on the reduced images. We concluded that ISOMAP (a nonlinear, local and geodesic method) was the most performing.

We extend here this study in comparing ISOMAP to 4 other DR methods that are Laplacien Eigenmaps, Second Order Blind Identification, Projection Pursuit and Sammon's Mapping. Our results show that Laplacian Eigenmaps, a nonlinear and local method like ISOMAP, but Euclidean, is the most interesting of all methods. The chapter is organized as follows: In Section 2, we briefly present the images we have processed, the reduction methods we have compared and the comparison procedure. Section 3 provides experimental results and Section 4 concludes the chapter.

6.2 Materials and Methods

Let $X = (x_1, \ldots, x_n)^T$ be the $n \times m$ data matrix. The number n represents the number of pixels in an image, and m the number of spectral bands. We have in our case $n=4096$ and $m=7$.

6.2.1 Satellite images

The images used in this study have been provided by the French Institute for the Environment (IFEN). They are multispectral images acquired by the LANDSAT 7 satellite in 2001. Each image consists of seven spectral bands ($m=7$), with a spatial resolution of 30 meters per pixel. The original images were cropped to 198 64×64 ($n=4096$) subimages, thus adapting the scale to our application, the study of avian territories. The rather small size of the subimages is also useful to limit the computational workload.

6.2.2 Estimating Intrinsic Dimensionality

One of our working hypotheses is that, though data points are points in $\mathbb{R}^m$, there exists a p-dimensional manifold $\mathcal{M}$ that can satisfyingly approximate the space spanned by the data points. The meaning of "satisfyingly" depends on the dimensionality reduction technique that is used. The so-called intrinsic dimension (ID) of X in $\mathbb{R}^m$ is the lowest possible value of p for which the approximation of X by $\mathcal{M}$ is reasonable. In order to estimate the ID of our data sets, we used a geometric approach that estimates the equivalent notion of fractal dimension [2]. Using this method, we estimated the intrinsic dimensionality of our dataset as being $p=3$.

6.2.3 Dimensionality Reduction

Dimensionality reduction techniques seek to represent X as a p-dimensional manifold ($p < m$) embedded in a m-dimensional space. DR methods can be classified according to three characteristics:

- ***Linear/nonlinear.*** This describes the type of transformation applied to the data matrix, mapping it from $\mathbb{R}^m$ to $\mathbb{R}^p$.

- ***Local/global.*** This reflects the kind of properties the transformation does preserve. In most nonlinear methods, there is a compromise to be made between the preservation of local topological relationships between data points, or of the global structure of X.

- ***Euclidean/geodesic.*** This defines the distance function used to estimate whether two data points are close to each other in $\mathbb{R}^m$, and should consequently remain close in $\mathbb{R}^p$, after the DR transformation.

ISOMAP. ISOMAP (Isometric Feature Mapping)[3] estimates the geodesic distance along the manifold using the shortest path in the nearest neighbors graph. It then looks for a low-dimensional representation that approximates those geodesic distances in the least square sense (which amounts to MDS). It consists of three steps:

1. Build $D_m(X)$, the all-pairs distance matrix.
2. Build a graph from X, using for each data point a restricted number of neighbors. For a given point x_i in $\mathbb{R}^m$, a neighbor is either one of the K nearest data points from x_i, or one for which $d_{ij}^m < \epsilon$. Build the all-pairs geodesic distance matrix $\Delta_m(X)$, using Dijkstras all-pairs shortest path algorithm.
3. Use classical MDS to find the transformation from $\mathbb{R}^m$ to $\mathbb{R}^p$ that minimizes

$$J_{isomap}(X,p) = \sum_{i,j=1}^{n} (\delta_{ij}^m - \delta_{ij}^p)^2 \tag{6.1}$$

ISOMAP is nonlinear, global and geodesic.

SOBI. Second-Order Blind Identification SOBI relies only on stationary second-order statistics that are based on a joint diagonalization of a set of covariance matrices. Each $X_i(t)$ is assumed to be an instantaneous linear mixture of n unknown components (sources) $s_i(t)$, via the unknown mixing matrix A.

$$X(t) = A_s(t) \tag{6.2}$$

This algorithm can be described by the following steps; more details on SOBI algorithm can be found in [4].

1. Estimate the sample covariance matrix $R_x(0)$ and compute the whitening matrix W with

$$R_x(0) = E(X(t)).X^*(t)) \tag{6.3}$$

2. Estimate the covariance matrices $R_z(\tau)$ of the whitened process $z(t)$ for fixed lag times τ.
3. Jointly diagonalize the set $\{R_z(\tau_j)/j = 1, \ldots, k\}$, by minimizing the criterion

$$J(M,V) = \sum_{k=1,\ldots,n} (\sum_{i \neq j=1,\ldots,n} |V^t M_{i,j} V|^2) \tag{6.4}$$

where M is a set of matrices in the form

$$M_k = V D_k V \tag{6.5}$$

where V is a unitary matrix, and D_k is a diagonal matrix.

4. Determine an estimation $\widehat{A}$ of the mixing matrix A such as

$$\widehat{A} = W^{-1} \tag{6.6}$$

5. Determine the source matrix and then extracting the p components.

SOBI is a linear, global, Euclidean method

Projection Pursuit. This projection method [5] is based on the optimization of the gradient descent. Our algorithm uses the Fast-ICA procedure that allows estimating the new components one by one by deflation. The symmetric decorrelation of the vectors at each iteration was replaced by a Gram-Schmidt orthogonalization procedure. When p components $w_1, \ldots, w_p$ have been estimated, the fix point algorithm determines w_{p+1}. After each iteration, the projections $w_{p+1}^T w_j w_j (j = 1, \ldots, p)$ of the p precedent estimated vectors are subtracted from w_{p+1}. Then w_{p+1} is renormalized:

$$w_{p+1} = w_{p+1} - \sum_{j=1}^{p} w_{p+1}^T w_j w_j \tag{6.7}$$

$$w_{p+1} = \frac{w_{p+1}}{\sqrt{w_{p+1}^T w_{p+1}}} \tag{6.8}$$

The algorithm stops when p components have been estimated.
Projection Pursuit is linear, global and Euclidean.

Sammon's Mapping. Sammon's mapping is a projection method that tries to preserve the topology of the set of data (neighborhood) in preserving distances between points [6]. To evaluate the preservation of the neighborhood topology, we use the following stress function:

$$E_{sammon} = \frac{1}{\sum_{i,j=1}^{N} d_{i,j}^n} \left(\sum_{i,j=1}^{N} \frac{(d_{i,j}^n - d_{i,j}^p)^2}{d_{i,j}^n} \right) \tag{6.9}$$

where $d_{i,j}^n$ and $d_{i,j}^p$ are the distances between points i^{th} and j^{th} points, in $\mathbb{R}^m$ and $\mathbb{R}^p$
This function, minimized by a gradient descent, allows adapting the distances in the projection space at best as distances in the initial space.

Sammons Mapping is a nonlinear, global, and Euclidean method.

Laplacian Eigenmaps. This method projects the set of data in a reduced subspace in preserving neighborhood relations between the points [7].
The three steps of the algorithm are the following:

1. Build the non-oriented symmetric neighborhood graph (each point is linked to its k nearest neighbors).
2. Associate a positive weight W_{ij} to each link of the graph. These weights can be constant ($W_{ij} = \frac{1}{k}$), or exponentially decreasing ($W_{ij} = e^{\frac{-\|x_i - x_j\|^2}{\sigma^2}}$).
3. Obtain the final coordinates y_i of the points in $\mathbb{R}^p$ by minimizing the cost function

$$E_L = \sum_{ij} \frac{W_{ij} \|y_i - y_j\|^2}{\sqrt{D_{ii}D_{jj}}} \tag{6.10}$$

where D is the diagonal matrix $D_{ii} = \sum_j W_{ij}$.
The minimum of the cost function is found with the eigenvectors of the Laplacian matrix:

$$L = I - D^{-\frac{1}{2}} - WD^{-\frac{1}{2}} \tag{6.11}$$

Laplacian Eigenmaps is a nonlinear, local, Euclidean method.

6.2.4 K-means Clustering

The K-means algorithm [8] is among the most popular and cost-effective clustering techniques. It finds the clustering result that minimizes the sum of squared Euclidean distances between data points and cluster centroids. To apply the K-means algorithm on images, we need to know the number of clusters of the image. In order to automate the process of our images, we added to the K-means algorithm a pre-processing that finds this number automatically. A description of this pre-processing can be found in [1].

6.2.5 Choice of Images and Manual Segmentation

All images were manually segmented by an expert in order to be compared with automatically segmented images. This manual segmentation is used as our "'golden standard"' to compare the clustering accuracy after DR.

6.2.6 Objective Comparison

The clustering results are compared to the expert segmentation using two objective measures: the correct classification rate, and the kappa statistic, computed from the confusion matrix [9].

6.3 Results

We have applied all DR methods on 20 Landsat images. Each reduction method provided a new 3 bands image.

Figure 1 shows one of the studied images. We can see here 2 views of it : a pseudo-color image created from the infrared bands of the original Landsat image and the classified image given by the expert.

Figure 2 shows the 3 bands images obtained with each reduction dimension method from the original image. There is no ordering of the different bands after dimensionality reduction. Note that Projection Pursuit provides a new image in which information is quiet deteriorated.

Figure 3 shows the classified images obtained from each reduced image. As noted in Figure 2, Projection Pursuit provides a classified image of very bad quality. Except for the Projection Pursuit, it's not possible to conclude from these images what method is better than the others.

Infrared 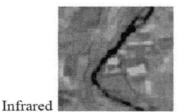  Expert

Fig. 6.1. Pseudo-color image created from the IR bands, and manual segmentation result.

Clustering results are summarized in Table 1. These values are averages of the values computed from the 20 tested images. For each image, we compared the automatic classified images with the expert classified image and computed the classification rate and the kappa value. We added in the table the run time of each DR program computed on a conventional desktop computer (P4 3GHz, 1GB RAM).

Observing these values, we can say that Sammon's Mapping (nonlinear, global, and Euclidean method) needs too much run-time for our application. SOBI and Projection Pursuit (linear, global, Euclidean methods) are very fast but the classification rates and kappa values of the images obtained with these methods are too low and don't allow us to recognize landscape organization. ISOMAP (nonlinear, global and geodesic method) and Laplacian Eigenmaps (nonlinear, local, Euclidean method) need acceptable run-times and provide images with sufficient information to obtain a correct classification. Between these two methods, Laplacain Eignemaps is the one which values are the most interesting.

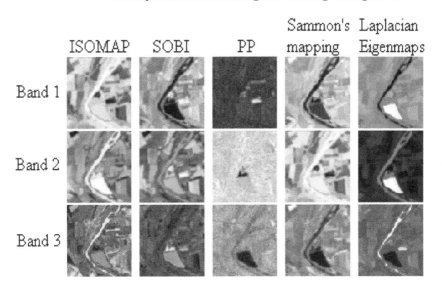

Fig. 6.2. Results of DR methods applied to a Landsat image.

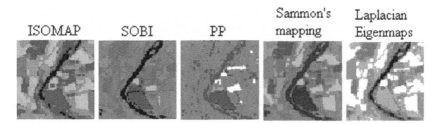

Fig. 6.3. Clustering results. The displayed colors correspond to the mean color for each cluster, computed from the infrared pseudo-color image from Fig. 2.

Table 6.1. Summary of clustering results for each DR method, showing run times, classification rates and kappa index for the confusion matrix.

Methods	Run Times (s)	Classification Rates(%)	Kappa
ISOMAP	81,05	34,24	0,77
SOBI	2,23	33,58	0,71
PP	2,40	23,42	0,42
Sammons mapping	2968,14	33,91	0,68
Laplacian Eigenmaps	28,61	38,87	0,78

6.4 Discussion

The comparison of DR algorithms has been studied mainly on synthetic data, by evaluating their ability to "unfold" a dataset. We took a more pragmatic approach, by measuring how DR can improve clustering results. Our results show that linear methods run faster than non-linear but they provide images with lower quality. Laplacian Eigenmaps provides the best rate "quality of image / run time". Projection Pursuit and Sammon's mapping are the less interesting methods because of the very poor quality of the provided images or the very high run-time. We are currently working on this study, in order to compare the data extracted from the classified reduced images in a program of landscape recognition.

Acknowledgments. The authors would like to thank Xavier Tizon, Jonathan Delcourt and Wail El Mjiyad for their helpful participation in this work.

References

1. Journaux, L., Tizon, X., Foucherot, I., Gouton, P.: Dimentionality Reduction Techniques : An Operational Comparison on Multispectral Satellite Images Using Unsupervised Clustering. IEEE 7th Nordic Signal Processing Symposium (NORSIG), Reykjavik, Iceland, June 7-9 (2006) 242-245.
2. Camastra, F., Vinciarelli, A.: Estimating the Intrinsic Dimension of Data with a Fractal-Based Method. IEEE Transactions on Pattern Analysis and Machine Intelligence, vol. 24 (2002) 1404-1407.
3. Tenenbaum, J. B., de Silva, V., Langford, J. C.: A Global Geometric Framework for Nonlinear Dimensionality Reduction. Science, vol. 290 (2000) 2319-2323
4. Belouchrani, A., Abed-Meraim, K., Cardoso, J. F., Moulines, E.: A blind source separation technique using second order statistics. IEEE Transactions on signal processing, vol. 45(2) (1997) 434-444.
5. Friedman, J. H., Tukey, J. W.: A projection pursuit algorithm for exploratory data analysis. IEEE Transactions on computers, vol. C23(9) (1974) 881-890.
6. Sammon, J. W.: A nonlinear mapping for data analysis. IEEE Transactions on Computers, vol. C-18 (1969) 401-409.
7. Belkin, M., Niyogi, P.: Laplacian eigenmaps for dimensionality reduction and data representation. Neural Computation, vol. 15(6) (2003) 1373-1396.
8. Duda, R. O., Hart, P. E., Stork, D. G.: Pattern Classification (2nd Edition). Wiley-InterScience (2001).
9. Tso, B., Mather, P. M.: Classification Methods for Remotely Sensed Data. Taylor and Francis Ltd (London) (2001).

7

SAR Image Compression based on Wedgelet-Wavelet

Ruchan Dong, Biao Hou, Shuang Wang, and Licheng Jiao

Institute of Intelligence Information Processing and National Key Lab for Radar Signal Processing, Xidian University 710071, Xian, China.
ruchandong@hotmail.com

Summary. Wavelet is well-suited for smooth images, but unable to economically represent edges. Wedgelet offers one powerful potential approach for describing edges in an image. it provides nearly-optimal representations of objects in the Horizon model, as measured by the minimax description length. In this paper, we proposed a multi-layered compression method based on Cartoon+Texture model which combined wedgelet and wavelet transforms, and where the coefficients of wedgelet and wavelet were coded with Huffman coding, run-length coding and SPIHT. Experiment results showed that the proposed method is effective and feasible in SAR image compression.

Key words: Wedgelet, Wavelet, SAR image compression, Cartoon + Texture model

7.1 Introduction

Synthetic aperture radar (SAR) imagesformed from spatially overlapped radar phase historiesare becoming increasingly important and abundant in a variety of remote-sensing and tactical application. With the increased abundance of these images, the need to compress SAR images without significant loss of perceptual image quality has become more urgent. There have been many algorithms for compressing SAR images, such as DCT transform [1], Gobar transform [2], and wavelet transform. Wavelet is more popular than any others, which has been used successfully by numerous authors for lossy image compression ([3], [4], and [5]).

Wavelet transform provides a sparse representation for smooth images, and is able to efficiently approximate smooth images. Unfortunately, this sparsity does not extend to piecewise smooth images, where edge disconti-nuities separating smooth regions persist along smooth contours. This lack of sparsity hampers the efficiency of wavelet based approximation and compression. In addition, wavelet decomposes low-frequency signal of image step by step, preserves lots of low-frequency information and discards high-frequency information, which leads to the loss of more high-frequency information. Wavelet coefficients in edges also have complex correlation. Ringing artifacts appear around edges when we reconstruct the SAR image.

Wedgelet developed by Donoho [6] offers one convenient approach for describing edges in an image. Each dyadic blocks of wedgelets, which contains a single straight edge with arbitrary orientation, can be chained together to approximate an edge contour with desired accuracy. For certain classes of contours, wedgelet had been shown to offer nearly-optimal nonlinear approximations. Unfortunately, due to errors introduced by wedgelet in the stage of approximating real edges, the application of wedgelet in SAR images compression is not straightforward.

We propose a novel method for SAR image compression in this paper. It combined wedgelet with wavelet transforms in SAR images compression [7], which takes advantage of the merits of wedgelet and wavelet. First, we could vertically collapse the edges in the image using wedgelet, and then it would be left with a texture image that could be efficiently compressed with wavelet. Moreover, since image edges tend to be smooth, they are low-dimensional structures that are easy to describe and compress explicitly (much like a smooth 1-d function can be compressed efficiently). So we can explain it by cartoon + texture ([7] [8]) model,

$$Image = \{cartoon\} + \{texture\}$$

$$f(x, y) = c(x, y) + t(x, y)$$

and get a two-layer compressing scheme, which first compresses the cartoon with wedgelet and then the texture residual regions by wavelet combined with SPIHT.

In Section 2, we introduce the mechanism of wedgelet from edgelet and explain the relationship of edgelets and wedgelet. More details of the method and encoding algorithms which are suitable to encode cartoon and texture regions are offered in Section 3. Experimental results are presented in section 4. Finally, we conclude in section 5 with a discussion.

7.2 The Mechanism of Wedgelet

Wedgelet was developed by Donoho [6] in 1997. It is a piecewise constant function on a dyadic square with a linear discontinuity, which can effectively represent edges of an image. From [6], it is clearly that wedgelet is composite of edgelets because of its decomposition correlated with edgelet. So we introduce edgelet firstly.

Edgelet and Edgelet Transform

Edgelet [6] is a finite dyadically-organizaed collection of line segments in the unit square, occupying a range of dyadic locations and scales, and occurring at a range of orientations. We would specifically describe the mechanism of edgelet.

Edgelet is constructed in a dyadic square. We define a dyadic square S [6][9] is the collection of points

$$\{(x_1, x_2) : [k_1/2^j, (k_1 + 1)/2^j] \times [k_2/2^j, (k_2 + 1)/2^j]\}$$

where $0 \leq k_1$, $k_2 \leq 2^j$ for an integer $j \geq 0$ .it can be written for clarity. Before introducing edgelet, we first have a concept of edgel.

We take the collection of all dyadic squares at scales $0 \leq j \leq J$. On each dyadic square, put equally-spaced vertices on the boundary, starting from corners, where equally-spaced is equal to δ, $\delta = 2^{-J-K}$ for $K \, \xi \, 0$. Label all the vertices in the clockwise boundary. Suppose we take vertices $v_1, v_2 \in [0, 1]^2$ and consider the line segment $e = \overline{v_1 v_2}$. We call such a segment to be an edgel (for edge element). With each dyadic square , edgelet is that the collection of all edgels connecting vertices on the boundary of . See Fig.1 for the inherent mechanism of edgelet [6].

Edgelets are not functions and do not make a basis; instead ,they can be viewed as geometric objects - line segments in the square. We can associate these line segments with linear functionals: for a line segment e and a smooth function $f(x_1, x_2)$, let $e[f] = \int_e f$ .Then the edgelet transform can be defined as

$$f(x) = \sum_{i_1 i_2} f(i_1, i_2) \phi_{i_1 i_2}(x) \tag{7.1}$$

$$T_f[e] = \int_e f = \int f(x(l)) dl \tag{7.2}$$

7.2.1 The Mechanism of Wedgelet

Edgelet can not make up a basis and has little approximation of image data, however, it offers a convenient description for wedgelet.

S is a dyadic square, if an edgelet $e \in E_\delta(S)$, where $E_\delta(S)$ is the collection of all edgelets, dose not lie entirely on a common edge of S, we say it is *nondegenerate* [6]. A nondegenerate edgelet traverses the interior of S, and splits S into two pieces, exactly one of which contains the segment of the boundary of S starting at $v_{0,s}$ and ending at $v_{1,s}$. Label the indicator of that piece $\omega_{\delta,S}$ and call this the wedgelet defined by e. Let $W_\delta(S) = \{1_S\} \bigcup \{\omega_{e,S}: e \in E_\delta(S) \text{ nondegenerate}\}$.

This collection of functions expresses all ways of splitting S into two pieces by edgelet splitting, including the special case of not splitting at all. Fig.2 gives us a more visualization to understand wedgelet and the relation between edgelet and wedgelet ([9], [10]).

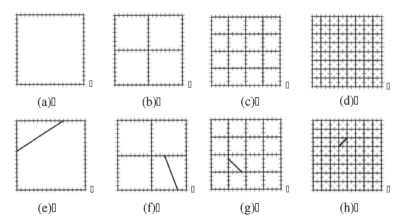

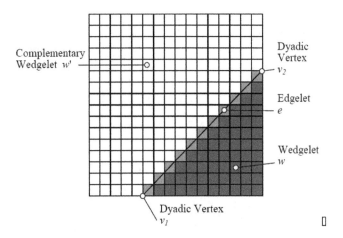

Fig. 7.1. Implement mechanism of edgelet, from (a) to (d) is marked out all vertices with spacing , connecting all pairs of vertices in the same square using line segments from (e) to (h)

Fig. 7.2. The relation between edgelet and wedgelet

7.3 SAR Image Compression via Wedgelet Based on Cartoon + Texture Model

SAR images differ from optical images, so we can't use the usual compression method to compress SAR im-ages. A SAR image includes lots of edges and texture regions, and there is a great deal of speckle noise and singularity in the texture regions. Standard wavelet algorithm can successfully compress texture regions of SAR images, but can't deal with the edge regions of SAR images. It is lucky that we find another tool wedgelet to compress the edges. Based on Cartoon + Texture model, we combined the two tools together for SAR image compression. ·

First, wedgelet was adopted to describe Cartoon regions of a SAR image. It is an efficient quantization method to make the coefficients integers. We used Huffman coding and run length coding to code these quanti-fied coefficients. Second, the texture regions of the SAR image ,which were got from original image subtracted by Cartoon regions, were compressed by combining wavelet and SPIHT algorithm.

7.3.1 Cartoon Regions Compression by Wedgelet

Wedgelet transform [9] and wedgelet decomposition can be written as following, Wedgelet transform:

$$W_f(\omega) \equiv \langle f, \omega \rangle, \forall \omega \in W \tag{7.3}$$

Wedgelet Decomposition:

$$f = \sum_{\omega \in W_{n,\delta}} a_\omega \omega \tag{7.4}$$

where using edgelet transform(reference to equation (1),(2))to compute wedgelet transform. Wedgelet transform is the process of wedgelet approximation which is a recursive dyadic adaptive partition ([6] [11]). It must meet the best condition of recursive dyadic partition, which is a complexity-penalized wedgelet partition.

$$f = \sum_{\omega \in W_{n,\delta}} a_\omega \omega \tag{7.5}$$

Where is optimal ED-RDP

$$min_{p \in EDRDP_{n,\delta}} \|y - AVE\{y|P\}\|^2 + \lambda \#P \tag{7.6}$$

ED-RDP means an edgelet-decorated recursive dyadic partition, let $AVE\{y|P\}$ be satisfy value of optimal recursive dyadic partition, $\lambda = \sqrt{4 \times \log(n)}\ \sigma$, we take σ to be the standard deviation of a SAR image.

Through wedgelet transform and decomposition for a SAR image, we got coefficients of cartoon regions. We used a simple and useful quantization method to make the coefficients integers after analyzing the coefficients and doing a number of experiments, these quantified coefficients were encoded with Huffman coding and run- length coding. We reconstructed cartoon regions of the SAR image after decoding, which is marked with $\widehat{c}(x,\ y)$.

7.3.2 Texture Regions Compression with Wavelet

After using wedgelet transform to approximate the SAR image, it was necessary to code the residual texture image $t = f - c$. The texture regions of SAR images carry useful information including thinner texture and singularity, an obvious alternative is simply to code the entire residual regions using wavelet. In this paper, we choose db9-7 [12] wavelet basis which is suitable for SAR images compression, adopted SPIHT [3] [4] to en-code wavelet coefficients. We obtained after decoding. We obtained $\widetilde{t}(x,y)$ after decoding.

Combined the two sections, we acquired the compression result of entire a SAR image, that was $\widehat{f}(x,y) = \widehat{c}(x,y) + \widehat{t}(x,y)$.

7.4 Experiments and Results Analysis

We choose two SAR images as test images which have abundant information in edges. The first image is a two-look X-Band amplitude (10 GHz) SAR called farm field in Bedfordshire, 3m resolution. The second image is a Ku-Band (15 GHz) SAR carried by the Sandia Twin Otter in California, named China Lake Airport, 3m resolu-tion. We contrasted our method with wavelet combined with SPIHT, where wavelet basis was db9-7. PSNR (peak-to-peak signal-to-noise ratio) measured the quality of SAR image compression

$$PSNR = 10lg\{\frac{255 \times 255}{\frac{1}{MN}\sum_{m=1}^{M}\sum_{n=1}^{N}[f(m,n) - \widehat{f}(m,n)]^2}\} \qquad (7.7)$$

where $f(m,n)$ is the original image, $\widehat{f}(m,n)$ is the reconstructed image.

Table 7.1. The compression performance of our method, gain in db reflects our method improvement over Wavelet. SAR image sizes: farm field (256×256), china lake airport (256×256)

SAR image	Rate (bpp)	Wavelet combined with SPIHT, PSNR(dB)	Our method PSNR(dB)	Gain (dB)
	0.25	20.5008	21.5332	1.0342
Farm Field	0.50	21.4420	22.4045	0.9625
	0.75	22.7138	23.5800	0.8662
	1.00	23.5753	24.4351	0.8598
China	0.25	23.4614	23.6859	0.2245
Lake	0.50	24.1424	24.5956	0.4532
Airport	0.75	25.1076	25.1893	0.0817
	1.00	25.6872	25.8414	0.1542

Table 1 shows compression results of wavelet and our method at a variety of bitrates for the two SAR images. In most cases, our method offers a higher PSNR than wavelet, with gain up to 1.0342dB.

The results of wavelet combined with SPIHT compression is shown in Fig.3 (b) and Fig.4 (b). It is obvious to find that strong ringing artifacts are introduced near edges, because the dependencies among wavelet coefficients are not properly exploited, quantization crashes the coherence. From Fig.3(c) and Fig.4 (c), the fact can be seen that the proposed method has more effective compression results with much less ringing around the dominant edges and better perceptual quality. Fig.5 shows three zooms on Fig.4 and one can see that our method performs very well in preserving more detail linear information of SAR images than wavelet.

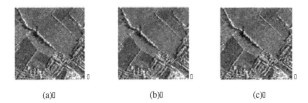

(a)   (b)   (c)

Fig. 7.3. Farm Field (a) Original image (25656). (b) The compression result of (a) using Wavelet with SPIHT, Rate=0.5bpp, PSNR=21.4420dB. (c) The Compression result of (a) Using our method, Rate=0.5bpp, PSNR =22.4045dB

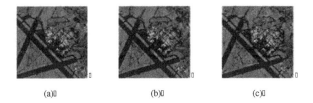

(a)   (b)   (c)

Fig. 7.4. Chinese Lake Airport (a) Original image(25656). (b)The compression result of (a) using Wavelet with SPIHT, Rate=0.5bpp, PSNR=24.1424dB. (c) The Compression result of (a) using our method, Rate=0.5bpp, PSNR =24.5956dB

7.5 Concluding Remarks

In this paper, we presented a novel method combined wedgelet with wavelet based on Cartoon + Texture model. Experiments showed that the proposed method is suitable for SAR images compression. Compared to wavelet, our method is more powerful for improving PSNR, decreasing ringing artifacts, as well as preferably preserving the thin regions of original SAR images. However, wedgelet uses adaptive quadtree partitioning, and unavoid-ably needs long computing time, which is a problem we will deal with in future.

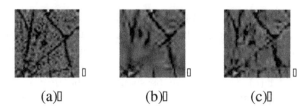

(a)   (b)   (c)

Fig. 7.5. Zooms on details of Fig.4. (a) Original: a zoom on local region of Fig.4 (a). (b) Wavelet combined with SPIHT: a zoom on local region of Fig.4 (b). (c) Our method: a zoom on local region of Fig.4 (c)

Acknowledgements.

This work is supported by the National Natural Science Foundation of China Grant No.60472084

References

1. Andreadis A, Benelli G, Garzelli A, Susini S.: A DCT-based adaptive compression algothrithm customized for radar imagery. IEEE Int, Geosciences and remote sensing-IGARSS, Vol. 4, Aug (1997) 1993-1995
2. Rober A. Baxter.: SAR image compressions with the Garbor transform. IEEE Trans, Geoscience and remote sensing, Vol. 37,No. 1, Jan (1999) 574-588
3. Shapiro J.: Embedded image coding using zerotrees of wavelet coefficients. IEEE Trans. Signal Processing, Vol. 41, Dec (1993) 3445-3462
4. Said, A Pearlman, W.A.: A new fast and efficient image codec based on set partitioning in hierarchical trees. IEEE Trans. Citcuits Syst. Video Techno, Vol. 6, June (1996) 243-250.
5. Brandfass M, Coster W, Benz U, Moreira A.: Wavelet Based approaches for efficient compression of complex SAR image data. IEEE Int, Geosciences and remote sensing-IGARSS, Vol. 4, Aug(1997) 2024-2027
6. Donoho D.L.: Wedgelets: nearly-minimax estimation of edges. Annals of Statistics, Vol. 27, (1999) 859-897
7. Wakin M, Romberg K, Choi. Hyekhochoi and Baraniuk R.: Image compression using an efficient edge cartoon + texture model. Proc, IEEE Date Compression conference-DCC. Snowbird, Utah, April (2002) 43-52
8. Michael B. Wakin M.: Image compression using multiscale geometric edge models. Master Thesis, Rice University, May (2002)
9. Donoho D.H, Huo Xiaoming.: Edgelts & Wedgelets, Software & Application. In KDI meeting. Oct (1999)
10. Huo Xiaoming.: Sparse image representation via combined transforms. PhD Thesis, Stanford University, Aug (1990)
11. Donoho D.H, Huo Xiaoming.: Beamlet pyramids: a new form of multiresolution analysis, suited for extraction lines, curves, and objects from very noisy image data. In Proceedings of SPIE, Vol. 4119, July (2000)

12. Antonini M, Barlaud M, Mathieu P and Daubechies I.: Image coding using wavelet transform. IEEE Tran. on Image Processing, Vol. 1, Issue.2, April (1992) 205-220

Part II

Texture Analysis and Feature Extraction

8

New approach of higher order textural parameters for image classification using statistical methods

Narcisse Talla Tankam[1,2], Albert Dipanda[1] and Emmanuel Tonye[2]

[1] Electronic, Computer science and Image Laboratory - LE2i -, Bourgogne University, Dijon, France[narcisse.talla, albert.dipanda]@u-bourgogne.fr
[2] Electronical and Signal processing Laboratory - LETS -, National Higher School Polytechnic, Yaounde, Cameroon tonyee@hotmail.com

Summary. Many researchers have demonstrated that textural data increase the precision of image classification when they are combined with grey level information. However, textural parameters of order two take too long computation time. The problem is more complex when one must compute higher order textural parameters, which however can considerably improve the precision of a classification. In this work, we propose a new formulation for the calculation of statistical textural parameters. The principle consists in reducing the calculation of a n-summation of type

$$\sum_{i_0=0}^{L-1}\sum_{i_1=0}^{L-1}\sum_{i_2=0}^{L-1}\cdots\sum_{i_{n-1}=0}^{L-1}\psi[i_0,i_1,\cdots,i_{(n-1)},P_{i_0},i_1,\cdots,i_{n-1}]$$

generally used in the evaluation of textural parameters, to a double summation of type $\sum_{p=0}^{W_x}\sum_{q=0}^{W_y}\chi(p,q)$ where L is the dynamic of grey levels (number of quantification levels) in the image, $(P_{i_0,i_1,\cdots,i_{(n-1)}})$ is the occurrence frequency matrix (co-occurrence matrix in the case of order two parameters) and W_x (respectively W_y) is the width (respectively the height) of the image window. This method produces the same results as the classical method, but it's about L^{n-1} times faster than the classical method and a gain of L^n of memory space is obtained, where n is the order of the textural parameter.

Key words: textural parameter, frequency matrix, SAR image, image classification

8.1 Introduction

A great number of studies have been carried out in order to highlight the perception of texture by the human eye. One can quote for example the psycho-visual experiments of B. Julesz [7] and the works of A. Gagalowicz

[4]. The texture analysis has been the object of many researches. Many methods has then been generated and developed in [2] and [5]. These methods aim to characterize, to describe, and to discriminate textures. The Grey Level Co-occurrence Matrix (GLCM) became the most used to classify satellite radar images texture [6],[13]. Several research axes have been developed in the field of GLCM. Some of them define a method to choose displacement vectors [11],[3], and others aim to simplify these matrices [15],[8].

Since the size of SAR images is rather significant, one of the priorities in the analysis of these images consists in the improvement of the computing time of texture parameters. Various authors leaned on this problem. M. Unser [14] proposed to replace the co-occurrence matrix by the sum and the difference of histograms which define the principal axes of the probabilities of second order stationary processes. D. Marceau et al. [10] proposed a textural and spectral approach for the classification of various topics and adopted the reduction of the level of quantification (16, 32 instead of 256), without deteriorating significantly the precision of the classification. A. Kourgly and A. Belhadj-Aissa [8] presented a new algorithm to calculate textural parameters using various histograms. This algorithm requires the allowance of a vector instead of a matrix, and the calculation of textural parameters is done according to new formulas by using a simple summation instead of a double summation, which reduces considerably the computing time. A. Akono et al. [1] proposed a new approach for the evaluation of the textural parameters of order 3, based on A. Kourgly and A. Belhadj-Aissa [8] works.

In this paper, we propose a new approach for the evaluation of textural parameters of any order $n \succ 1$. This method consists in a reformulation of textural parameters, which avoids the hard calculation of the occurrence frequency matrix (GLCM in the case of order 2).

The reminder of this paper is organized as follows. Section 2 presents the classical textural parameter formulation. In Section 3, we describe the proposed textural parameter reformulation. Section 4 is devoted to an example of the calculation of a very used texture parameter using both classical and new methods. In Section 5, a comparative study is done on the algorithmic complexity required by each approache for the evaluation of a textural parameter. Some experimental results are included in Section 6 and the conclusion follows.

8.2 Classical formulation of statistical textural parameters

Basically, statistical textural parameters are function of the occurrence frequency matrix (OFM), which is used to define the occurrence frequency of each n-ordered grey levels $(i_0, i_1, \cdots, i_{n-1})$ that follow a certain condition called 'connection rule' in an image window.

8.2.1 The occurrence frequency matrix (OFM)

In an image with L levels of quantification, the OFM of order $n \succ 1$ is a L^n size matrix. In this matrix, each element $(P_{i_0,i_1,i_2,\cdots,i_{n-1}})$ expresses the occurrence frequency of the n-ordered pixels $(i_0, i_1, \cdots, i_{n-1})$ following the connection rule $R_n(d_1, d_2, \cdots, d_{n-1}, \theta_1, \theta_2, \cdots, \theta_{n-1})$. This connection rule defines the spatial constraint that must be verified by the various positions of the n-ordered pixels $(i_0, i_1, \cdots, i_{n-1})$ used for the evaluation of the OFM. This rule means that the pixel i_{k+1} with $(0 \prec k and k \prec n)$ is separated to the pixel i_k by d_{k-1} pixel(s) in the θ_k direction. For the sake of simplicity, $R_n(d_1, d_2, \cdots, d_{n-1}, \theta_1, \theta_2, \cdots, \theta_{n-1})$ will be noted by R_n in the following.

8.2.2 Textural parameters

Let's consider an image window F of size $NL \times NC$, where NL is the number of lines and NC is the number of columns. The classical expression of textural parameters concerned by this approach is given by the following expression.

$$Para_n = \sum_{i_0=0}^{L-1} \sum_{i_1=0}^{L-1} \cdots \sum_{i_{n-1}=0}^{L-1} [\psi(i_0, i_1, \cdots, i_{n-1}) \times P_{i_0,i_1,\cdots,i_{n-1}}] \quad (8.1)$$

where $(P_{i_0,i_1,\cdots,i_{n-1}})$ is the OFM, n is the order of the parameter and Ψ is a real function defined by $\psi : IN^n \longrightarrow IR$
The following table 8.4 gives examples of some used textural parameters.

Table 8.1. examples of some used statistical textural parameters

Parameter	Formulation			$\psi(i_0, \cdots, i_{n-1})$
Mean	$\sum_{i_0=0}^{L-1} \sum_{i_1=0}^{L-1} \cdots \sum_{i_{n-1}=0}^{L-1} i_0 \times P_{i_0,\cdots,i_{n-1}}$			i_0
Dissymetry	$\sum_{i_0=0}^{L-1} \sum_{i_1=0}^{L-1} \cdots \sum_{i_{n-1}=0}^{L-1}$	$\sum_{k=0}^{n-1} \sum_{l=k+1}^{n} \|i_k - i_l\| \times P_{i_0,\cdots,i_{n-1}}$		$\sum_{k=0}^{n-1} \sum_{l=k+1}^{n-1} \|i_k - i_l\|$
Inv. Diff.	$\sum_{i_0=0}^{L-1} \sum_{i_1=0}^{L-1} \cdots \sum_{i_{n-1}=0}^{L-1}$	$\dfrac{P_{i_0,\cdots,i_{n-1}}}{1 + \sum_{k=0}^{n-1} \sum_{l=k+1}^{n} \|i_k - i_l\|}$		$\dfrac{1}{1 + \sum_{k=0}^{n-1} \sum_{l=k+1}^{n} \|i_k - i_l\|}$

8.3 New formulation

The goal is to avoid the calculation of both $P_{i_0,i_1,\cdots,i_{n-1}}$ and the n-summation $\sum_{i_0=0}^{L-1} \sum_{i_1=0}^{L-1} \cdots \sum_{i_{n-1}=0}^{L-1}$ which are very expensive in computing time and memory space.

Since it is true that:

$$a \times b = \underbrace{a + a + \ldots + a}_{b \ times} \tag{8.2}$$

Equation 8.1 can be written in the following form:

$$Para_n = \sum_{i_0=0}^{L-1} \sum_{i_1=0}^{L-1} \cdots \sum_{i_{n-1}=0}^{L-1} \underbrace{\psi(i_0, i_1, \cdots, i_{n-1}) + \cdots + \psi(i_0, i_1, \cdots, i_{n-1})}_{P_{i_0,i_1,\cdots,i_{n-1}} \ times}$$

$$\tag{8.3}$$

During the image window scanning, each time that a n-ordered pixels $(i_0, i_1, \cdots, i_{n-1})$ that follows the connection rule R_n is obtained, the value of $\psi(i_0, i_1, \cdots, i_{n-1})$ is calculated and stored. At the end of the image window scanning, the sum of all the stored values is equal to the product $\psi(i_0, i_1, \cdots, i_{n-1}) \times P_{i_0,i_1,\cdots,i_{n-1}}$. This operation discards the used of the term $P_{i_0,i_1,\cdots,i_{n-1}}$ as expected. Let's now show how to discard the used of the expression $\sum_{i_0=0}^{L-1} \sum_{i_1=0}^{L-1} \cdots \sum_{i_{n-1}=0}^{L-1}$. Indeed, Equation 8.3 can be written as the following.

$$Para_n = \begin{array}{l} \underbrace{\psi(i_0^1, i_1^1, \cdots, i_{n-1}^1) + \cdots + \psi(i_0^1, i_1^1, \cdots, i_{n-1}^1)}_{P_{i_0,i_1,\cdots,i_{n-1}} \ times} \\ + \quad \underbrace{\psi(i_0^2, i_1^2, \cdots, i_{n-1}^2) + \ldots + \psi(i_0^2, i_1^2, \cdots, i_{n-1}^2)}_{P_{i_0,i_1,\cdots,i_{n-1}} \ times} \\ + \quad \cdots \\ + \underbrace{\psi(i_0^m, i_1^m, \cdots, i_{n-1}^m) + \ldots + \psi(i_0^m, i_1^m, \cdots, i_{n-1}^m)}_{P_{i_0,i_1,\cdots,i_{n-1}} \ times} \end{array} \quad m = L^n. \tag{8.4}$$

Equation 8.4 is equivalent to the following.

$$Para_n = \sum_{(i_0,i_1,\cdots,i_{n-1})} \psi(i_0, i_1, \cdots, i_{n-1}) \left[\delta \langle (i_0, i_1, \cdots, i_{n-1}), R_n, F \rangle \right]_0^1 \tag{8.5}$$

where $\left[\delta \langle (i_0, i_1, \cdots, i_{n-1}), R_n, F \rangle \right]_0^1$ is a boolean function that takes the value '1' if $(i_0, i_2, \cdots, i_{n-1})$ follows the connection rule R_n and '0' otherwise. Equation 8.5 can then be written as the following.

$$Para_n = \sum_{(i_0,i_1,\cdots,i_{n-1})R_n,F} \psi(i_0, i_1, \cdots, i_{n-1}) \tag{8.6}$$

where $(i_0, i_1, \cdots, i_{n-1})_{R_n, F}$ is the set of all the n-ordered pixels $(i_0, i_1, \cdots, i_{n-1})$ in the image window F that follow the connection rule R_n. When the connection rule is defined, from a given pixel in the window F, there is at most one n-ordered pixels $(i_0, i_1, \cdots, i_{n-1})$ that verifies the connection rule R_n. Thus, the set $(i_0, i_1, \cdots, i_{n-1})_{R_n, F}$ is obtained by scanning the window F at once. Let's now define a function φ_{R_n} as follows.

$$\varphi_{R_n} : \begin{aligned} IN^2 &\longrightarrow IN^n \\ (p, q) &\longmapsto (i_0, i_1, \cdots, i_{n-1}) \end{aligned} \tag{8.7}$$

where i_0 is the pixel at the position (p,q) in the image window F and $(i_0, i_1, \cdots, i_{n-1})$ is the n-ordered pixels that the respective positions in the image window follow the connection rule R_n. The substitution of equation 8.7 in equation 8.6 gives the following equation.

$$Para_n = \sum_{p=0}^{NC-1} \sum_{q=0}^{NL-1} \psi\left(\varphi_{R_n}(p, q)\right) \left[\delta_{R_n}(p, q, F)\right]_0^1 \tag{8.8}$$

where NC and NL are respectively the number of columns and the number of lines of the image window F; $\delta_{R_n}(p, q, F)$ is a boolean function that takes the value '1' if from the position (p,q) in the image window F, it can be obtained a n-ordered pixel positions in F following the connection rule R_n. It takes the value '0' otherwise. In practice, this function contributes to determine the variation domain of p and q. So, equation 8.8 can also be written as the following.

$$Para_n = \sum_{p=Tb_x}^{Te_x} \sum_{q=Tb_y}^{Te_y} \left\{\psi\left(\varphi_{R_n}(p, q)\right)\right\} \tag{8.9}$$
$$with 0 \leq Tb_x \leq Te_x \prec NC and 0 \leq Tb_y \leq Te_y \prec NL$$

Equation 8.8 or equation 8.9 are the new formulation of statistical textural parameters. In the following, we present an example to assess the equivalence between the new and the classical formulation. It's important to mention that the proposed formulation can be parallelized while computing. Therefore, We are now able to do classification using statistical approach in real time independently to the image sizes.

8.4 Calculation of the dissymmetry parameter

This section helps to understand our methodology. Let's consider the image window F_1 illustrated by Figure 1. We have randomly chosen to calculate the dissymmetry parameter, using both the classical and the proposed approaches, with the connection rule $R_2(2, 45°)$.

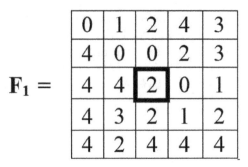

Fig. 8.1. experimental image window

8.4.1 Classical approach

At the order 2, the classical formulation of the dissymmetry parameter is given by the following equation.

$$Diss_2 = \sum_{i=0}^{4}\sum_{j=0}^{4}|i - j|P_{ij} \qquad (8.10)$$

This approach uses the GLCM P_2, which is given by the following equation.

$$P_2 = \begin{pmatrix} 0\,0\,0\,0\,0 \\ 0\,0\,0\,0\,0 \\ 1\,0\,0\,2\,0 \\ 0\,0\,1\,0\,0 \\ 1\,1\,2\,0\,1 \end{pmatrix} \qquad (8.11)$$

The evaluation of the dissymmetry parameter, according to equation 10 is done as follows.

$$
\begin{aligned}
Diss_2 = {}&|0-0|.P_{00} \;+|0-1|.P_{01} \;+|0-2|.P_{02} \;+|0-3|.P_{03} \;+|0-4|.P_{04} \\
&+|1-0|.P_{10} \;+|1-1|.P_{11} \;+|1-2|.P_{12} \;+|1-3|.P_{13} \;+|1-4|.P_{14} \\
&+|2-0|.P_{20} \;+|2-1|.P_{21} \;+|2-2|.P_{22} \;+|2-3|.P_{23} \;+|2-4|.P_{24} \\
&+|3-0|.P_{30} \;+|3-1|.P_{31} \;+|3-2|.P_{32} \;+|3-3|.P_{33} \;+|3-4|.P_{34} \\
&+|4-0|.P_{40} \;+|4-1|.P_{41} \;+|4-2|.P_{42} \;+|4-3|.P_{43} \;+|4-4|.P_{44}
\end{aligned}
$$

So $Diss_2 = \mathbf{16}.$

8.4.2 New approach

From equations 8.1 and 8.10, one can deduce that for each couple of grey levels (i,j) that the positions follow the connection rule $R_2(2, 45^o)$ in the window F_1, we have:

$$\psi(i,j) \quad = \quad |i-j| \tag{8.12}$$

In Figure 8.2, the connection rule $R_2(2, 45°)$ is materialised in the image window F_1 by an arrow. Its origin (a ring) materialises the position of the pixel 'i' and its end (a dotted line square with rounded angles) materialises the position of the pixel 'j'. A couple of grey levels (i,j) follows the connection rule if and only if the position of 'i' (respectively 'j') is in the ring (respectively the dotted line square) of the same arrow. Figure 2 enumerates the set of all couples (i,j) of grey levels that follow the connection rule R_2 in the image window F_1. There is a total number of 9 couples (i,j) of grey levels following the connection rule in the image window F_1

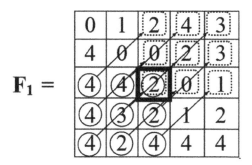

Fig. 8.2. : Illustration of the connection rule on the image window F_1

One can now determine the domain of variation of p and q in equation 8.9. This domain is given by the following equation.

$$\left\{ \begin{array}{ll} Tb_x = 0 & \qquad Tb_y = 2 \\ Te_x = 2 & \qquad Te_y = 4 \end{array} \right\} \tag{8.13}$$

From equations 8.9 and 8.13, one can evaluate the various values of the function φ_{R_n} as follows.

$|\varphi_{R_n}|2$	$|3$	$|4$	
$|0$	$|(4,2)$	$|(4,0)$	$|(4,2)$	
$|1$	$|(4,4)$	$|(3,2)$	$|(2,0)$	
$|2$	$|(2,3)$	$|(2,3)$	$|(4,1)$	

The possible values of q (2,3,4) are set in columns and the possible values of p (0,1,2) are set in lines.
The evaluation of the dissymmetry parameter, using the new approach is then

done as follows.

$$
\begin{aligned}
DISS_2 &= \psi(\varphi_{R_n}(0,2)) & &+\psi(\varphi_{R_n}(0,3)) & &+\psi(\varphi_{R_n}(0,4)) \\
&+\psi(\varphi_{R_n}(1,2)) & &+\psi(\varphi_{R_n}(1,3)) & &+\psi(\varphi_{R_n}(1,4)) \\
&+\psi(\varphi_{R_n}(2,2)) & &+\psi(\varphi_{R_n}(2,3)) & &+\psi(\varphi_{R_n}(2,4)) \\
&= \psi(4,2) & &+\psi(4,0) & &+\psi(4,2) \\
&+\psi(4,4) & &+\psi(3,2) & &+\psi(2,0) \\
&+\psi(2,3) & &+\psi(2,3) & &+\psi(4,1) \\
&= |4-2| & &+|4-0| & &+|4-2| \\
&= |4-4| & &+|3-2| & &+|2-0| \\
&= |2-3| & &+|2-3| & &+|4-1|
\end{aligned}
$$

$$= 16$$

This example confirms that the new and the classical approaches are equivalent as expected. Let's now evaluate the algorithmic complexity of the two approaches.

8.5 Complexity evaluation

Let's evaluate the calculation complexity and the storage requirement of each approach. Let's consider an image F of NL lines and NC columns, and L the number of grey level quantification in the image. Then let's evaluate and compare the respective operational and the spatial complexities of the two approaches.

8.5.1 Classical approach

Let's consider $N(\psi)$ the required number of operations for the evaluation of $\psi(i_0, i_1, \cdots, i_{n-1})$ in equation 2 for a given n-ordered $(i_0, i_1, \cdots, i_{n-1})$ pixels. This number is given by the following equation 8.14.

$$N(\psi) = O(n) \tag{8.14}$$

The evaluation of the OFM requires a number N(P) of operations given by the following equation.

$$N(P) = o(NL \times NC) \tag{8.15}$$

Knowing that the term $\sum_{i_0=0}^{L-1} \sum_{i_1=0}^{L-1} \cdots \sum_{i_{n-1}=0}^{L-1}$ in equation 1 implies that there is exactly L^n different possible n-ordered pixels $(i_0, i_1, \cdots, i_{n-1})$, the evaluation of the textural parameter by the classical approach requires a number $N_{classic}(Para_n)$ of operations given by the following equation.

$$N_{classic}(Para_n) = O(n) \times L^n + o(NL \times NC) \quad (8.16)$$

8.5.2 New approach

Tall For each couple (p,q) of values in the definition domain, the number of operations required for the evaluation of $\psi[\varphi_{R_n}(p,q)]$ is given in equation 9 because $\varphi_{R_n}(p,q)$ replaces only the n-ordered pixels $(i_0, i_1, \cdots, i_{n-1})$ following the connection rule R_n and where the first pixel i_0 occupies the position (p,q) in the image window F. According to equation 8.9, the number Nc of couples (p,q) required for the evaluation of the textural parameter is given by the following equation.

$$N_c = o(NL \times NC) \quad (8.17)$$

The required number $N_{new}(Para_n)$ of operations for the evaluation of the textural parameter $Para_n$, using the new approach is then given by the following equation.

$$N_{new}(Para_n) = O(n) \times o(NL \times NC) \quad (8.18)$$

8.5.3 Comparison of the two approaches

The following equation gives the ratio between claculated complexity while using the classical approach and the claculated complexity while using the new approach for the evaluation of a textural parameter.

$$R = (N_{classic}(Para_n)/N_{new}(Para_n)) = O(L^{n-1}) \quad (8.19)$$

With the aim to illustrate the previous equation 19, let's evaluate the parameter 'Mean' in the various orders, for an image window F of size 25×25. As connection rule, let's consider $R_2(1,0^o)$ (for order 2), $R_3(1,1,0^o,0^o)$ (for order 3), $R_4(1,1,1,0^o,0^o,0^o)$ (for order 4) and $R_5(1,1,1,1,0^o,0^o,0^o,0^o)$ (for order 5) .
The classical formulation of the 'Mean' parameter for an order n $(2 \leq n \leq 5)$ with 256 levels of quantifications (L=256) is given by the following equation.

$$\mu_{i_0} = \sum_{i_0=0}^{2} 55 \sum_{i_1=0}^{2} 55 \cdots \sum_{i_{n-1}=0}^{2} 55 \left(i_0 \times P(i_0, i_1, \cdots, i_{n-1}) \right) \quad (8.20)$$

With the new formulation, the same parameter is given by the following equation.

$$Mean_n = \sum_{p=0}^{25-n} \sum_{q=0}^{24} \psi \left[\varphi_{R_n}(p,q) \right] \quad (8.21)$$

with $\varphi_{R_n}(p,q)$ given by the following equation.

$$\varphi_{R_n}(p,q) = (F(p,q), F(p+1,q), \cdots, F(p+(n-1),q)) \tag{8.22}$$

where F(i,j) is the pixel at the position (i,j) in the image window F.

Table 8.5 presents the calculated complexity (number of operations) required by each approach for the evaluation of the parameter 'Mean', for the various orders n = 2, 3, 4, 5, using respectively the connection rules $R_2(1,0^o)$, $R_3(1,1,0^o,0^o)$, $R_4(1,1,1,0^o,0^o,0^o)$ and $R_5(1,1,1,1,0^o,0^o,0^o,0^o)$, on the experimental image window F1.

Table 8.2. Comparison of the calculated complexities required for the evaluation of the parameter Mean, using the two approaches

Order	Classical Approach (CA)	Proposed Approach (PA)	$R = \frac{CA}{PA}$	$\tau = \frac{R}{L^{n-1}}$
2	$6,614 \times 10^4$	600	110,22	0,430
3	$1,678 \times 10^7$	575	110,22	0,444
4	$4,29 \times 10^9$	550	110,22	0,447
5	$1,1 \times 10^{12}$	525	110,22	0,487

According to Table 8.5, one can notice that the proposed formulation is about $\tau \times L^{n-1}$ times more economic than the classical formulation in term of calculated complexity (number of operations required), with $\tau \approx 1/2$ in the case of the parameter 'Mean'.

8.6 Storage requirement

Although the memory space is no longer a critical problem today, the higher level of memory space gain obtained by this new approach convinces us to present a comparative study.

8.6.1 Classical approach

In the classical approach, the evaluation of textural parameters requires the evaluation of the OFM. For a given order n, the evaluation of the OFM requires a memory space given by the following equation.

$$S(P) = L^n \tag{8.23}$$

The image storage requires a memory space given by the following equation.

$$S(I) = NL \times NC \tag{8.24}$$

The memory space required for the evaluation of the textural parameter with the classical approach is then given by the following equation.

$$S_{classic} = S(P) + S(I) = L^n + NL \times NC \tag{8.25}$$

8.6.2 New approach

With the new approach, the OFM is not evaluated. The computer memory space required for the evaluation of the textural parameter is only the space memory required to store the image, given by the following equation.

$$S_{new} = NL \times NC \tag{8.26}$$

8.6.3 Comparison of the two approaches

From equation 8.25 and equation 8.26 one can deduce following equation.

$$S_{classic} = L^n + S_{new} \tag{8.27}$$

From equation 8.27, one can notice that L^n is the memory space difference required by the classical and the new approaches for the evaluation of an order n textural parameter. This quantity is considerable. In fact, for an image with 256 levels of quantification, at the order 5, L^n gives 2^{40}.

8.7 Experimental results

For experimentation, let's consider the image of size 500×500 in Figure 8.3. This image is obtained by concatenation of 4 Brodatz textures.

In this section, we present for each approach, the required complexity for the evaluation of each textural parameter presented in table 1 and we make a comparative study of the obtained results.

8.7.1 Calculated complexity

Figure 8.4. presents the required number of operations for the evaluation of each parameter specified in table 1, using the classical and the new approaches for the various orders. For the evaluation, we considered the connection rules $R_2(1,0^o)$, $R_3(1,1,0^o,0^o)$, $R_4(1,1,1,0^o,0^o,0^o)$ and $R_5(1,1,1,1,0^o,0^o,0^o,0^o)$ respectively for orders 2, 3, 4 and 5. We used image window of size 5x5.

Fig. 8.3. experimental image

Mean		
Order	Classic	New
2	1,31E+5	100000
3	3,35E+7	75000
4	8,59E+9	50000
5	2,19E+12	25000

Dissymmetry		
Order	Classical	New
2	4,91E+8	150000
3	3,77E+11	337500
4	1,93E+14	450000
5	7,14E+16	325000

Inverse Difference		
Order	Classic	New
2	8,19E+08	250000
3	4,61E+11	412500
4	2,14E+14	500000
5	7,69E+16	350000

Fig. 8.4. Experimental results for the parameter 'Dissymmetry'

8.7.2 Experimentation complexity

Let's now evaluate the required time for the evaluation of the various parameters specified in table 1. The image in figure 1 has been used as experimental image. For each parameter, we evaluate the required time, using separately the new and the classical approaches, at the various orders. For each order, the evaluation is done for three different window size (5×5, 9×9 and 11×11). As experimentation environment, we used a computer with AMD Athlon processor of 706 MHz of speed and 128 Mo of RAM and we obtained the following results.

The 'Mean' parameter

Figure 8.5 below presents the required time variation for the evaluation of the 'Mean' parameter

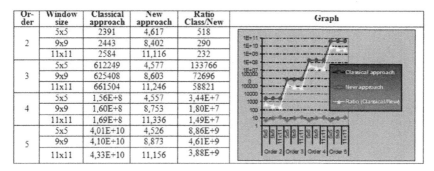

Order	Window size	Classical approach	New approach	Ratio Class/New	Graph
2	5x5	2391	4,617	518	
	9x9	2443	8,402	290	
	11x11	2584	11,116	232	
3	5x5	612249	4,577	133766	
	9x9	625408	8,603	72696	
	11x11	661504	11,246	58821	
4	5x5	1,56E+8	4,557	3,44E+7	
	9x9	1,60E+8	8,753	1,80E+7	
	11x11	1,69E+8	11,336	1,49E+7	
5	5x5	4,01E+10	4,526	8,86E+9	
	9x9	4,10E+10	8,873	4,61E+9	
	11x11	4,33E+10	11,156	3,88E+9	

Fig. 8.5. Experimental results for the parameter 'Mean'

The 'Dissymmetry' parameter

Figure 8.6 below presents the required time variation for the evaluation of the 'Dissymmetry' parameter

Order	Window size	Classical approach	New approach	Ratio Class/New	Graph
2	5x5	2414,91	4,766	506,69	
	9x9	2467,43	8,933	276,21	
	11x11	2609,84	16,453	158,62	
3	5x5	618371,49	4,987	123996,68	
	9x9	631662,08	9,854	64102,09	
	11x11	668119,04	14,55	45918,83	
4	5x5	1,57E+8	4,909	3,20E+7	
	9x9	1,61E+8	10,685	1,51E+7	
	11x11	1,70E+8	13,089	1,30E+7	
5	5x5	4,05E+10	4,977	8,13E+9	
	9x9	4,14E+10	11,717	3,53E+9	
	11x11	4,37E+10	11,757	3,72E+9	

Fig. 8.6. Experimental results for the parameter 'Dissymmetry'

The 'Inverse Difference' parameter

Figure 8.7 below presents the required time variation for the evaluation of the 'Inverse Difference' parameter

8.7.3 Discussion

According to the various results above, one can notice that the required time for a texture parameter evaluation does not necessarily increase with the window size. Using the classical approach, the required time exponentially increases with the parameter order. But using the new approach, the required

Or-der	Window size	Classical approach	New approach	Ratio Class/New	Graph
2	5x5	2414,91	4,766	506,69	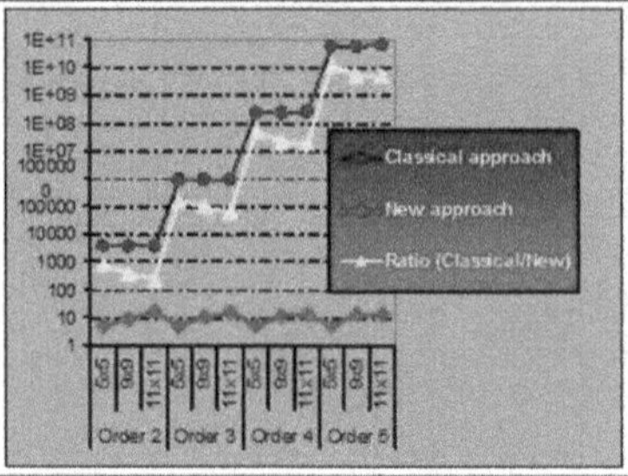
	9x9	2467,43	8,933	276,21	
	11x11	2609,84	16,453	158,62	
3	5x5	6,18E+5	4,987	1,23E+5	
	9x9	6,31E+5	9,854	6,41E+4	
	11x11	6,68E+5	14,55	45918,83	
4	5x5	1,57E+8	4,909	3,20E+7	
	9x9	1,61E+8	10,685	1,51E+7	
	11x11	1,70E+8	13,089	1,30E+7	
5	5x5	4,05E+10	4,977	8,13E+9	
	9x9	4,14E+10	11,717	3,53E+9	
	11x11	4,37E+10	11,757	3,71E+9	

Fig. 8.7. Experimental results for the parameter 'Inverse Difference'

time doesn't significantly change. Talla et al. in [12] showed that a combination of various orders of texture parameters gives a better result in image classification, while compared to the texture classification using the neighboring gray level dependence matrix method [11]. The above experimental results confirm the theoretical result and show that, using the new approach, a combination of several texture parameters for texture classification requires less time than using the neighboring gray level dependence matrix method. Another important advantage of this approach is that, there is no need of reducing the number of gray level in the image [10] before computing the texture parameter.

8.8 Conclusion

Statistical methods of image classification having proven reliable, the main problem on which are confronted the researchers were how to reduce the computing time of the frequency matrix required. The goal of this work was to propose a new method of evaluation of statistical textural parameters without calculating the OFM. The proposed method does not take into consideration textural parameters, which are expressed as a quadratic or logarithmic function of the OFM. To rich this end, a new formulation of textural parameters has been developed, discarding the used of the OFM. With the new approach, the number of computing operations required for the evaluation of textural parameter is reduced about $\tau \times L^{n-1}$ times and a gain of L^n of computer memory space is obtained. In contrary to the neighboring gray level dependence matrix for texture classification, the new approach can be parallelized while computing. Image classification using statistical approach can now be done in real time independently to the image size.

References

1. Akono, A., Tonye, E., Nyoungui A., N. (2003) Nouvelle méthodologie d'évaluation des paramètres de texture d'ordre trois. International Journal of Remote Sensing, vol 24, N 9, pages 1957-1967

2. Conners, R.W. and Harlow, C.A. (1980), A theoretical comparison of texture algorithms. PAMI, 2(3) :204-222, April 1980.

3. Davis, L.S., Johns,S. and Aggarwal, J.K. (1979) Texture analysis using generalized co-occurrence matrices. PAMI, 1(3) :251-259, July 1979

4. Gagalowicz, A. (1980) Visual discrimination of stochastic texture fields based upon their second order statistics. In: ICPR80. pp 786-788.

5. Haralick, R.M. (1979), Statistical and structural approaches to texture. PIEEE, 67(5) :786-804, May 1979.

6. Haralick, R.M., Shanmugam, K. and Dinstein, I. (1973) Textural features for image classification. SMC, 3(6) :610-621, November, 1973.

7. Julesz, B. (1962) Visual pattern discrimination. IT, 8(2):84-92, February 1962.

8. Kourgly, A., and Belhadj-Aissa, A. (1999), Nouvel algorithme de calcul des paramtres de texture applique a la classification d'images satellitaires. Actes des 8emes Journees Scientifiques du Reseau Teledetection de l'AUF, Lausanne, 22 au 25 novembre 1999.

9. Lohmann, G. (1995) Analysis and synthesis of textures: a cooccurrence-based approach. CG, 19 :29-36.

10. Marceau, D., Howarth, P. J., Dubois, J. M. (1990) and Gratton, D. J., Evaluation of the grey-level co-occurrence method for land-cover classification using SPOT imagery. IEEE Transactions on Geosciences and Remote Sensing, 28, 513-519.

11. Sun, C. and Wee, W.G. (1983) Neighboring gray level dependence matrix for texture classification. CVGIP, 23:341-352.

12. Talla, T., Dipanda, A., Tonye, E., and Akono, A. (2006) Méthode optimisée de classification d'images radar RSO par usage des paramètres de texture d'ordre élevé. Application à la mangrove littorale Camerounaise. Proceedings of the 8th African colloquium on the research in computer science (CARI), Cotonou, Benin.

13. Terzopoulos, D., and Zucker, S.W. (1980) Texture discrimination using intensity and edge co-occurrences. In ICPR80, pages 565-567.

14. Unser, M. 1986 Sum and difference histogram for texture classification. IEEE Transactions on Pattern Analysis and Machine Intelligence, 8, pp 118-125.

15. Zucker, S.W. and D. Terzopoulos (1980) Finding structure in co-occurrence matrices for texture analysis. CGIP, 12(3) :286-308.

Texture Discrimination Using Hierarchical Complex Networks

Thomas Chalumeau[1], Luciano da F. Costa[2], Olivier Laligant[1], and Fabrice Meriaudeau[1]

[1] Universite de Bourgogne - Le2i, 12 rue de la Fonderie 71200 Le Creusot - France `t.chalumeau@iutlecreusot.u-bourgogne.fr`
[2] Universidade de Sao Paulo- IFSC, Caixa Postal 369 - CEP 13560-970 - Sao Carlos - SP - Brasil `luciano@if.sc.usp.br`

Summary. Texture analysis represents one of the main areas in image processing and computer vision. The current article describes how complex networks have been used in order to represent and characterized textures. More specifically, networks are derived from the texture images by expressing pixels as network nodes and similarities between pixels as network edges. Then, measurements such as the node degree, strengths and clustering coefficient are used in order to quantify properties of the connectivity and topology of the analyzed networks. Because such properties are directly related to the structure of the respective texture images, they can be used as features for characterizing and classifying textures. The latter possibility is illustrated with respect to images of textures, DNA chaos game, and faces. The possibility of using the network representations as a subsidy for DNA characterization is also discussed in this work.

9.1 Introduction

Textures are everywhere: in nature as well as in human-made objects and environments. As such, texture provides important information from which to identify objects and also to infer physical properties of scenes (e.g. gradients of texture may indicate scene depth). Although the difference between textures and other images (i.e. involving objects and shapes) remains unclear, such a decision is often important because the methods applied in image processing and analysis often differ depending on the type of images (i.e. texture against shape analysis). Interestingly, the issues of texture definition and representation/characterization are therefore intensely intertwined.

The continuing investigations in texture have considered several alternative approaches such as the Fourier and wavelet transform, co-occurrence matrices and derived measurements. Generally, textures are characterized by a high degree of disorder and/or periodical information, or hybrids of these two principles. The presence of periodicity is closely related to spatial correlations,

while large levels of disorder are associated to high entropy and lack of correlations. Because these two opposing features often co-exist in textures, it
is important to consider methods for representation and analysis of textures
which can be capable of accounting for these two effects.

Introduced recently [1, 2, 3], complex networks can be conceptualized as
an interface between graph theory and statistical physics, two traditional and
well-established research areas. Basically, complex networks are characterized
by the presence of patterns of connections which are different from those
observed in regular networks (e.g. lattices) or even random networks. Two
important manifestations of such "complexity" are the small world property,
namely the fact that a pair of the network nodes tend to be interconnected
through a short path, and the scale free property, indicating that the distribution of nodes is scale invariant, implying the presence of hubs. Interestingly,
the complexity in such networks is also characterized by the co-existence of
local and global features, in analogous fashion to the organization of textures.
For instance, even random networks will present more densely connected sub-
structures (the so-called communities) as a consequence of statistical fluctua-
tions. Such an inherent representational ability of complex networks has been
explored in order to represent and analyze textures and images [4, 5]. One of
the simplest ways to represent textures as complex networks is by expressing
pixels as nodes and similarity between the gray-level or local features of the
texture image as edges. Then, complex networks measurements (see [2]) can
be obtained so as to provide an objective quantification of the properties of
the texture in terms of the topological and connectivity of the respectively
obtained network. Hierarchical extensions of these measurements [6], which
consider further neighborhoods around each node, can also be used in order to provide additional information about the textures [7]. In particular,
the ability of such hierarchical features to express from local (i.e. close node
neighborhoods) to global (i.e. more distant neighborhoods) properties of the
texture contributes further to integrating the local and global aspects often
found in images. In addition to illustrating the possibility to use the framework described above for image/texture classification considering a database
of real textures and some hierarchical measurements (considering parameters
different from those presented in [7], the current article also discusses the important issue of using complex network representations as a subsidy for analysis of DNA sequences obtained through chaos games (e.g. [8]), which are also
considered as part of the image database adopted in the present work. This
article starts by briefly presenting complex networks and their measurements,
and follows by illustrating the suggested approach to texture classification
considering traditional images as well as DNA images.

9.2 Complex Networks

As a graph, complex networks involve a set of nodes and a set of edges between such nodes. Edges can be binary (i.e. presence or absence of connection) or weighted, and directed or not. The present work is limited to non-directed edges. Complex networks are typically represented by a matrix W such that $W(j, i)$ is the weight of the edge extending from node i to node j, with $i, j = 1, 2, ..., N$. The characterization of the topological and connectivity properties of complex networks can be achieved by using measurements borrowed from graph theory (e.g. [9]) and complex network research (e.g. [3]) including but being by no means limited to:

9.2.1 Degree and Strength Distributions

The degree of a given node is equal to the number of connections which it makes. For weighted connections the degree of a node is called strength and corresponds to the sum of all the weights of the respective links. The frequency histograms of the degrees (or strengths), as well as the respectively inferred average values, provide an important characterization of the connectivity of the network under analysis. In particular, scale free networks are characterized by histograms following a straight line when represented in log-log axes.

9.2.2 Clustering Coefficient

The clustering coefficient of a given node i is defined as:

$$C_i = \frac{Number\ of\ connections\ between\ nodes\ connected\ to\ node\ i}{Number\ of\ possible\ connections\ between\ these\ nodes} \quad (9.1)$$

whenever the denominator is equal to zero, we impose $C_i = 0$. Note that it follows that $0 < C_i < 1$ for any possible node. Figure 9.1 illustrates the calculation of the clustering coefficient for a simple network.

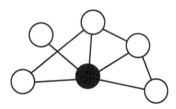

Fig. 9.1. Calculation of the clustering coefficient C_i of the node represented in black: $C = \frac{3}{5 \times 4 \backslash 2} = \frac{3}{10}$.

9.2.3 Hierarchical measurements

Several complex networks measurements, including the node degree and clustering coefficient, can be generalized to take into account not only the immediate neighborhood of a node, but also those which are at successive distances (i.e. 2, 3, ...) from that specific node [6]. In particular, the hierarchical degree of a node for hierarchical level i corresponds to the number of edges connecting the nodes at distance i to the nodes at distance $i + 1$. The hierarchical clustering coefficient of a given node for hierarchical level i is calculated in the same way as the traditional clustering measurement, but considering the edges between the nodes at distance i and the nodes at distance $i + 1$.

9.3 Image Representation As Complex Networks

For simplicity's sake, the images are assumed to be a square with $M \times M$ pixels. The transformation of an image into a graph [4, 7] considered in this work involves the representation of each pixel as a node and similarities between the gray levels (or other local properties such as color and depth) of pairs of pixels as the weights of the respective edges. The so obtained complex network can therefore be represented by a weight matrix W with dimension $M^2 \times M^2$. It is also interesting to obtain a matrix W_t containing only the connections defined by higher similarities between the pixels (i.e. only the elements of W which are smaller than a threshold T are kept). The differences between gray levels (which define the edge weights) are calculated only inside a circular region of radius r centered at each pixel. Such a procedure avoids border effects (the pixels near the border are less connected) and also imposes spatial constraints (adjacency) on the obtained complex network. Figure 9.2 illustrates the representation of a simple image as a complex network including or not the borders.

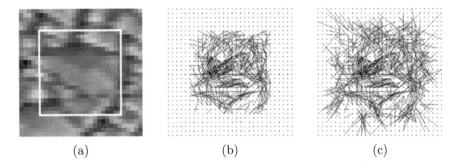

(a)   (b)   (c)

Fig. 9.2. (a): 30×30 sub-image with the 20×20 utile zone, (b): typical representation of the complex network with *threshold* $= \pm 2$, (c): representation without the border effect problem.

9.4 Texture Characterization

The characterization of the texture properties of the considered images can be accomplished by using complex networks measurements such as the node degree (traditional and hierarchical), strength and clustering coefficient (traditional and hierarchical). These measurements can be obtained from the matrix Wt by using the methodology described in [6], which involves a growing wavefront emanating from the reference node in order to determine the successive hierarchical levels. Two lists are used: current and next, so that the latter contains the neighbors of the nodes in the former, to be visited at the next step, after which the lists are updated. The average values (first moment) of each of these measurements are considered in the present work. We consider three types of images: three categories of textures obtained from the CUReT[3] database; images of faces; and images of DNA virus sequences from the EMBL[4] site, obtained through chaos game (see below). Each of these categories is represented by 10 respective samples, and all images are of size 30×30. The DNA images were obtained by using the chaos game representation [10, 8]. Basically, a DNA sequence is composed by a stream of four bases a, c, g, and t, such as $cggtggca$.... The image space into which such a sequence is to be represented has a base assigned to each of its corners: $a = (0,0)$; $c = (1,0)$; $g = (1,1)$ and $t = (0,1)$. The initial base is represented as a point at the middle of the respective quadrant, and for each new base, the image pixel laying in the middle of the position of the previous point and the corner identified by the current base is incremented of 1. The considered DNA images were obtained from virus sequences with 10000 nucleotides. Figure 9.3 shows one of the DNA images considered in this work.

Figure 9.4 illustrates the scatterplot obtained by considering discriminant analysis (e.g. [11]) and all the 5 types of images. More specifically, the images were translated into respective graphs and the average node degree, strength and clustering coefficients were calculated considering only the first hierarchical level (additional levels are considered in the next section of this article). This set of measurement was then projected into two dimensions so as to maximize the separation between the 5 classes, as quantified by the inter and intra-class dispersions [12].

It is clear from this result that each of the types of images yielded a well-defined respective cluster, with relatively little overlap between one another. Different dispersions were obtained for each cluster, with the faces subset of images implying the largest scattering and the DNA images yielding a relatively compact cluster. The overlaps between some of the clusters are largely a consequence of the projection of the data onto two-dimensions and can be minimized by considering further hierarchical levels.

---

[3] http://www1.cs.columbia.edu/CAVE/software/curet/
[4] http://www.ebi.ac.uk/embl/

Fig. 9.3. Image obtained from a virus DNA sequence with 10000 bases by using the chaos game approach.

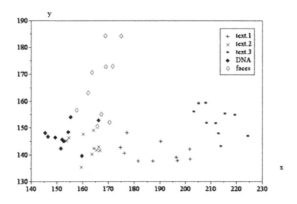

Fig. 9.4. The scatterplot obtained by canonical variable analysis considering the first hierarchical levels of the node degree, clustering coefficient and strength. The two main canonical variables, which are linear combinations of the considered measurements, are represented as x and y.

9.5 The Effect Of Hierarchies On Classification

While the above results were obtained by using the measurements of degree, strength and clustering coefficient considering only the first hierarchical level, enhanced information about the spatial organization of the images can be obtained by using further hierarchies. This possibility is illustrated in this section with respect to the same database considered in the previous section. The software Tanagra [13], involving a multilayer perceptron, has been applied in order to separate the image classes. The adopted perceptron has 25 neurons and its dynamics was limited to a maximum of 500 iterations. The learning rate was 0.25 and the error rate threshold was 0.01. The network is fed with all images and respective classes (training) and then requested to classify the same data. The quality of the classification is quantified in terms of the quantity.

The ratio of classification was calculated for all hierarchical levels, i.e. from 1 to 6. Note that the maximum limit of the possible hierarchical levels is implied by the fact that the considered complex networks are finite, so that the maximum level correspond to their respective diameters (i.e. the maximum path length between any two nodes). The evolution of the correct classification ratio in terms of the hierarchical levels, represented along the x-axis, is given in Figure 9.5. Note that the ratio value obtained for a specific hierarchical level k considers the measurements obtained for all hierarchial levels up to k, and not just that hierarchical level k. Three evolutions of the classification ratio are shown in this figure, each considering a different threshold value used for translating the image into the respective complex network.

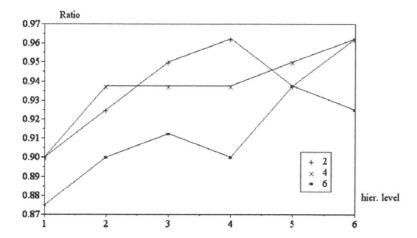

Fig. 9.5. Ratio of correct classification in terms of the hierarchical levels considering thresholds $T = 2$, 4 and 6.

It is clear from these results that the use of measurements considering progressive hierarchical levels has a definite effect in increasing the classification ratio. Interestingly, the ratio obtained for threshold 2 presents a peak, falling along the fifth and sixth hierarchical levels. This indicates that the respectively threshold network, the less dense in connections among the three considered cases, has relatively few shortest paths extending further than 4 edges.

9.6 Concluding Remarks

The present article has shown how to represent, characterize and classify textures by using complex networks. Measurements of the topology and connectivity of these networks, especially when considered in their hierarchical versions, have an inherent ability to represent and characterize the co-existing local and global features often present in textures, resulting particularly interesting for texture analysis and classification. The potential of such an approach has been illustrated with respect to a texture database including faces and images obtained from DNA sequences by using chaos games. The promising results obtained regarding the separation of such classes corroborates the use of the complex network approach to identify between DNA images obtained from coding and non-coding sequences, an important problem in genetics. The possibility to use complex networks representations and respective measurements as a subsidy to texture definition and characterization provides another interesting prospect for further investigations.

References

1. Mark E. J. Newman, "The structure and function of complex networks," *cond-mat/0303516*, 2003.
2. Luciano da Fontoura Costa, Francisco A. Rodrigues, Gonzalo Travieso, and P. R. Villas Boas, "Characterization of complex networks: A survey of measurements," *cond-mat/0505185*, 2005.
3. Reka Albert and Albert-Laszlo Barabasi, "Statistical mechanics of complex networks," *cond-mat/0106096*, 2001.
4. Luciano da Fontoura Costa, "Complex networks, simple vision," *cond-mat/0403346*, 2004.
5. Luciano da Fontoura Costa, "Hub-based community finding," *cond-mat/0405022*, 2004.
6. Luciano da Fontoura Costa, "Hierarchical characterization of complex networks," *cond-mat/0412761*, 2005.
7. T. Chalumeau, L. da F. Costa, O. Laligant, and F. Meriaudeau, "Optimized texture classification by using hierarchical complex networks measurements," *Machine Vision Applications in Industrial Inspection XIV*, vol. 6070, 2006.
8. J. S. Almeida, J. A. Carriço, A. Maretzek, P. A. Noble, and M Fletcher, "Analysis of genomic sequences by chaos game representation," *BIOINFORMATICS*, vol. 17, pp. 429–437, 2001.
9. Reinhard Diestel, *Graph Theory*, Springer, 2nd electronic edition edition, 2000.
10. H. Joel Jeffrey, "Chaos game representation of gene structure," *Nucleic Acids Research*, vol. 18 (8), pp. 2163–2170, 1990.
11. R. O. Duda, P. E. Hart, , and D. G. Stork, "Pattern classification," *Wiley Interscience*, 2001.
12. Luciano da Fontoura Costa and Roberto M. Cesar Junior, *Shape Analysis and Classification: Theory and Practice*, CRC Press, Inc, 2000.
13. Ricco Rakotomalala, "Tanagra : un logiciel gratuit pour l'enseignement et la recherche," *in Actes de EGC'2005, RNTI-E-3*, vol. 2, pp. 697–702, 2005.

Error analysis of subpixel edge localisation

Patrick Mikulastik, Raphael Höver and Onay Urfalioglu

Information Technology Laboratory (LFI), Leibniz University of Hannover

Summary. In this work we show analytically and in real world experiments that an often used method for estimating subpixel edge positions in digital camera images generates a biased estimate of the edge position. The influence of this bias is as great as the uncertainty of edge positions due to camera noise. Many algorithms in computer vision rely on edge positions as input data. Some consider an uncertainty of the position due to camera noise. These algorithms can benefit from our calculation by adding our bias to their uncertainty.

Key words: Subpixel accuracy, Edge detection, Parabolic, Regression

10.1 Introduction

The low level task of precise edge detection is a basis for many applications in image processing and computer vision.

In this work we show analytically and in real world experiments that an often used method for estimating subpixel edge positions in digital images generates a biased estimate of the edge position. We show analytically that common algorithms similar to the well known Canny [1] edge detector, which use parabolic functions for subpixel refinement of edge positions exhibit a bias of the estimage.

So far extensive work has been done on edge localisation, as it is a crucial task in computer vision and image processing. Nevertheless we think that we can make some additions to this toppic. The effect of the bias revealed by our calculation is in the same range as the uncertainty of edge positions introduced by camera noise. Our theoretical results are verified in easily reproducible experiments with real world data.

Some past approaches to edge localisation [2, 3, 4, 5] have shown good or even optimal approaches for continuous signals. Nevertheless, they are not easily portable to the case of digitised images because it is the interpolation process that introduces the bias to the edges positions. Other approaches [1, 6, 7] use digitised images but apply no subpixel interpolation of edge positions.

Kisworo et. al. [8] apply a local energy approach to localise subpixel edges. However their experiments don't show if there is a systematic bias for their approach. Rockett et. al. [9] and Mikulastik [10] examine parabolic interpola- tion for subpixel edge localisation. Mikulastik focuses on dealing with camera noise and Rockett et. al. [9] find that there is no bias for edges parallel to the sampling raster. This is because they use synthetic images for fitting their edge model and no real world images. However, none of the researches men- tioned above has measured or described the error that leads to a systematic bias, that we are introducing here.

In the next section we introduce our signal model and describe a sample edge detector similar to the detector by introduced by Canny [1]. Afterwards in section 10.3 we show in a calculation that a parabolic interpolation generates a bias, which we also measure in CCD camera images in section 10.4. Finally we give some conclusions and a summary in section 10.5.

10.2 Edge Detection

The following steps lead to an edge detector similar to the Canny [1] edge detector. We assume that the pulse response of the image acquisition system, in our case a CCD camera, can be modelled by a Gaussian with variance σ_{cam}. This is widely accepted and applied in the literature [11, 2]. We model the edges to be localised by a Gaussian with variance σ_{edge} convolved with an ideal step. An ideal step edge modelled this way would have a variance of $\sigma_{edge} = 0$. With $I(x)$ being the intensity at a coordinate x, the signal for an edge at position x_{max} can be modelled with

$$I(x) = \int_x (h_{edge}(x - x_{max}) * h_{cam}(x)) \, dx$$

with

$$h(x) = -\frac{x}{\sqrt{2\pi\sigma^3}} \cdot e^{-\frac{1}{2}\left(\frac{x}{\sigma}\right)^2} \quad . \tag{10.1}$$

The following approach describes a 1D filter applied in horizontal direction, for detection of mostly vertical edges. It can be applied in two passes in horizontal and vertical direction, so that all edges in a 2D image can be detected.

To be able to detect edges as maxima we have to generate a gradient image. Since an ideal gradient operator would introduce aliasing to the image we need a lowpass filter combined with a gradient. We choose the first derivate of a Gaussian. The Gaussian acts as lowpass filter. Combined with the first derivate we get the desired gradient. Choosing an *impulse invariance design* approach we sample the analog derivate of the Gaussian to get our digital filter. We get minimal distortion through aliasing and windowing with $\sigma_{grad} = 1.0062$ for a five tab filter.

Because of the gradient operation our edges now have the form of sampled Gaussian functions. The sampled signal $g(x)$ we get after application of the gradient filter described above is:

$$\frac{\partial}{\partial x}I(x) = g(x) = (h_{edge}(x - x_{max}) * h_{cam}(x) * h_{grad}(x))$$
$$= (h_{all}(x - x_{max})) \quad ,$$

where the functions $h(x)$ have the same form as seen in equation 10.1. The variance of the Gaussians is:

$$\sigma_{all} = \sqrt{\tilde{\sigma}^2_{edge} + \tilde{\sigma}^2_{cam} + \sigma^2_{grad}}$$
$$= \sqrt{\left(\frac{\sigma_{edge}}{\cos(\varphi)}\right)^2 + \left(\frac{\sigma_{cam}}{\cos(\varphi)}\right)^2 + \sigma^2_{grad}} \quad . \tag{10.2}$$

The angle φ is zero for edges orthogonal to the filter and greater for rotated edges.

The generation of the gradient signal is very similar to that used in the well known detector by Canny [1]. Many approaches now use regression with a parabolic function to find the subpixel peak points in the gradient image that give us the exact edge positions. The following calculation shows that this leads to a systematic bias.

10.3 Calculation of bias

We can write the following equation for a parabola

$$\hat{g}_{par}(x) = a \cdot x^2 + b \cdot x + c \quad . \tag{10.3}$$

The hat on $\hat{g}_{par}(x)$ indicates an estimated value for the real value $g(x)$, which is not available. The maximum of the parabola lies at

$$\hat{x}_{max} = -\frac{b}{2a} \quad . \tag{10.4}$$

Figure 10.1 shows an example of a parabola fitted to the signal $g(x)$. From the figure we can deduce that the parabola approximates the signal $g(x)$ only in a window with the width w_{fit}. Outside this window the differences between parabola and $g(x)$ become quite big. Furthermore the signal $g(x)$ does not have the form of a parabola. It is just approximated by it. Therefore the maximum value x_{max} exhibits a systematic bias to the real maximum of $g(x)$. In the following section we show the derivation the systematic error $e_{sys} = \hat{x}_{max} - x_{max}$.

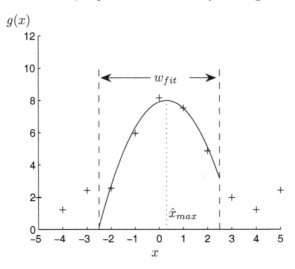

Fig. 10.1. A parabola fitted to the signal $g(x)$. It is fitted to $w_{fit} = 5$ samples. The further away the sample positions are from $\hat{x}_{max}$ the more the samples of the gradient signal differ from a parabolic interpolation.

10.3.1 Systematic error

The systematic error is a function of the window width w_{fit} and the parameters σ_{all} and x_{max}.

$$e_{sys} = e_{sys}\left(w_{fit}, \sigma_{all}, x_{max}\right) \tag{10.5}$$

It is the difference between the estimated value $\hat{x}_{max}$ and the real value x_{max}.

$$
\begin{aligned}
e_{sys}\left(w_{fit}, \sigma_{all}, x_{max}\right) &= \hat{x}_{max} - x_{max} \\
&= -\frac{b}{2a} - x_{max}
\end{aligned}
\tag{10.6}
$$

For regression with a parabolic function, as described in eq. 10.3, the following equation is valid and can be solved for a, b and c:

$$
\begin{pmatrix} x_1^2 & x_1 & 1 \\ \vdots & \vdots & \vdots \\ x_{wfit}^2 & x_{wfit} & 1 \end{pmatrix} \cdot \begin{pmatrix} a \\ b \\ c \end{pmatrix} = \begin{pmatrix} g_{par}(x_1) \\ \vdots \\ g_{par}(x_{wfit}) \end{pmatrix}
\tag{10.7}
$$

$$
\underline{A} \cdot \begin{pmatrix} a \\ b \\ c \end{pmatrix} = \begin{pmatrix} g_{par}(x_1) \\ \vdots \\ g_{par}(x_{wfit}) \end{pmatrix}
\tag{10.8}
$$

$$\begin{pmatrix} a \\ b \\ c \end{pmatrix} = (\underline{A}^T \underline{A})^{-1} \underline{A}^T \begin{pmatrix} g_{par}(x_1) \\ \vdots \\ g_{par}(x_{wfit}) \end{pmatrix}$$

$$= \underline{M} \cdot \begin{pmatrix} h_{all}(x_1 - x_{max}) \\ \vdots \\ h_{all}(x_{wfit} - x_{max}) \end{pmatrix}$$

$$= \underline{M} \cdot \mathbf{h}_{all}(\mathbf{x}_{wfit} - \mathbf{x}_{max}) \qquad (10.9)$$

with the matrix

$$\underline{M} = \begin{pmatrix} \mathbf{m}_1^T \\ \mathbf{m}_2^T \\ \mathbf{m}_3^T \end{pmatrix} = (\underline{A}^T \underline{A})^{-1} \underline{A}^T \qquad (10.10)$$

and the vectors

$$\mathbf{h}_{all}(\mathbf{x}) = \begin{pmatrix} h_{all}(x_1) \\ \vdots \\ h_{all}(x_N) \end{pmatrix} \quad \mathbf{x}_{wfit} = \begin{pmatrix} x_1 \\ \vdots \\ x_{wfit} \end{pmatrix} \quad \mathbf{x}_{max} = \begin{pmatrix} x_{max} \\ \vdots \\ x_{max} \end{pmatrix} \qquad (10.11)$$

With these expressions eq. 10.6 can be written as

$$e_{sys}(w_{fit}, \sigma_{all}, x_{max}) = -\frac{b}{2a} - x_{max}$$

$$= -\frac{1}{2} \frac{\mathbf{m}_2^T \cdot \mathbf{h}_{G,Ges}(\mathbf{x}_{wfit} - \mathbf{x}_{max})}{\mathbf{m}_1^T \cdot \mathbf{h}_{G,Ges}(\mathbf{x}_{wfit} - \mathbf{x}_{max})} - x_{max} \qquad (10.12)$$

For the following calculation it is assumed that the maximal sample of $g(x)$ is located at position $x = 0$. This doesn't limit generality, since it can be achieved by a simple coordinate transformation. Therefore we can write $\underline{A}$ as:

$$\underline{A} = \begin{pmatrix} \left(-\frac{w_{fit}-1}{2}\right)^2 & -\frac{w_{fit}-1}{2} & 1 \\ \vdots & \vdots & \vdots \\ \left(\frac{w_{fit}-1}{2}\right)^2 & \frac{w_{fit}-1}{2} & 1 \end{pmatrix} . \qquad (10.13)$$

The elements $d_{i,j}$ of a matrix $\underline{D} = \underline{A}^T \underline{A}$ are:

$$d_{1,1} = \left(-\frac{w_{fit}-1}{2}\right)^4 + \left(-\frac{w_{fit}-1}{2}+1\right)^4 + \cdots + \left(-\frac{w_{fit}-1}{2}+w_{fit}-1\right)^4$$

$$= \sum_{k=0}^{w_{fit}-1}\left(-\frac{w_{fit}-1-2k}{2}\right)^4 = 2\cdot\sum_{k=0}^{\frac{w_{fit}-1}{2}} k^4 \tag{10.14}$$

$$d_{1,2} = \left(-\frac{w_{fit}-1}{2}\right)^3 + \left(-\frac{w_{fit}-1}{2}+1\right)^3 + \cdots + \left(-\frac{w_{fit}-1}{2}+w_{fit}-1\right)^3$$

$$= \sum_{k=0}^{w_{fit}-1}\left(-\frac{w_{fit}-1-2k}{2}\right)^3 = 0 \tag{10.15}$$

$$d_{1,3} = \left(-\frac{w_{fit}-1}{2}\right)^2 + \left(-\frac{w_{fit}-1}{2}+1\right)^2 + \cdots + \left(-\frac{w_{fit}-1}{2}+w_{fit}-1\right)^2$$

$$= \sum_{k=0}^{w_{fit}-1}\left(-\frac{w_{fit}-1-2k}{2}\right)^2 = 2\cdot\sum_{k=0}^{\frac{w_{fit}-1}{2}} k^2 \tag{10.16}$$

and

$$d_{2,1} = 0 \qquad d_{2,2} = d_{1,3} \qquad d_{2,3} = 0 \tag{10.17}$$
$$d_{3,1} = d_{1,3} \qquad d_{3,2} = 0 \qquad d_{3,3} = w_{fit} \tag{10.18}$$

and the abbreviations μ and λ defined as:

$$\mu = 2\cdot\sum_{k=0}^{\frac{w_{fit}-1}{2}} k^4 \tag{10.19}$$

$$\lambda = 2\cdot\sum_{k=0}^{\frac{w_{fit}-1}{2}} k^2 \quad, \tag{10.20}$$

the matrix $\underline{D} = \underline{A}^T\underline{A}$ can be written as:

$$\underline{D} = \begin{pmatrix} \mu & 0 & \lambda \\ 0 & \lambda & 0 \\ \lambda & 0 & w_{fit} \end{pmatrix} \quad. \tag{10.21}$$

The inverse $\underline{D}^{-1}$ is:

$$\underline{D}^{-1} = \begin{pmatrix} \frac{1}{\mu}-\frac{\lambda^2}{\mu(\lambda^2-\mu w_{fit})} & 0 & \frac{\lambda}{\lambda^2-\mu w_{fit}} \\ 0 & \frac{1}{\lambda} & 0 \\ \frac{\lambda}{\lambda^2-\mu w_{fit}} & 0 & -\frac{\mu}{\lambda^2-\mu w_{fit}} \end{pmatrix} \quad. \tag{10.22}$$

Now, the matrix $\underline{M} = (\underline{A}^T\underline{A})^{-1}\underline{A}^T$ becomes:

$$\underline{M} = \begin{pmatrix} \frac{1}{\mu} - \frac{\lambda^2}{\mu(\lambda^2-\mu w_{fit})} & 0 & \frac{\lambda}{\lambda^2-\mu w_{fit}} \\ 0 & \frac{1}{\lambda} & 0 \\ \frac{\lambda}{\lambda^2-\mu w_{fit}} & 0 & -\frac{\mu}{\lambda^2-\mu w_{fit}} \end{pmatrix} \cdot \begin{pmatrix} \left(-\frac{w_{fit}-1}{2}\right)^2 & \cdots & \left(\frac{w_{fit}-1}{2}\right)^2 \\ -\frac{w_{fit}-1}{2} & \cdots & \frac{w_{fit}-1}{2} \\ 1 & \cdots & 1 \end{pmatrix}$$

$$= \begin{pmatrix} \frac{\lambda-\frac{w_{fit}}{4}(w_{fit}-1)^2}{\lambda^2-\mu w_{fit}} & \frac{\lambda-\frac{w_{fit}}{4}(w_{fit}-3)^2}{\lambda^2-\mu w_{fit}} & \cdots & \frac{\lambda-\frac{w_{fit}}{4}(w_{fit}-2w_{fit}+1)^2}{\lambda^2-\mu w_{fit}} \\ \frac{1}{\lambda}\cdot\left(-\frac{w_{fit}-1}{2}\right) & \cdots & \cdots & \frac{1}{\lambda}\cdot\left(\frac{w_{fit}-1}{2}\right) \\ \frac{\frac{\lambda}{4}(w_{fit}-1)^2-\mu}{\lambda^2-\mu w_{fit}} & \frac{\frac{\lambda}{4}(w_{fit}-3)^2-\mu}{\lambda^2-\mu w_{fit}} & \cdots & \frac{\frac{\lambda}{4}(w_{fit}-2w_{fit}+1)^2-\mu}{\lambda^2-\mu w_{fit}} \end{pmatrix} \quad (10.23)$$

with the vectors $\mathbf{m}_1^T$ and $\mathbf{m}_2^T$:

$$\mathbf{m}_1 = \frac{1}{\lambda^2-\mu w_{fit}} \cdot \begin{pmatrix} \lambda - w_{fit}\left(\frac{w_{fit}-1}{2} - 0\right)^2 \\ \lambda - w_{fit}\left(\frac{w_{fit}-1}{2} - 1\right)^2 \\ \vdots \\ \lambda - w_{fit}\left(\frac{w_{fit}-1}{2} - (w_{fit}-1)\right)^2 \end{pmatrix} \quad (10.24)$$

$$\mathbf{m}_2 = \frac{1}{\lambda} \cdot \begin{pmatrix} -\frac{w_{fit}-1}{2} \\ \vdots \\ \frac{w_{fit}-1}{2} \end{pmatrix} \quad . \quad (10.25)$$

As an example we consider an ideal edge with $\sigma_{edge} = 0$ and $\varphi = 0$. For the camera we choose $\sigma_{cam} = 1$, and for the gradient filter we set $\sigma_{grad} = 1,0062$ as chosen in section 10.2. For this example σ_{all} is

$$\sigma_{all} = \sqrt{\left(\frac{0}{\cos(0)}\right)^2 + \left(\frac{1}{\cos(0)}\right)^2 + 1,0062^2} = 1.4186 \quad . \quad (10.26)$$

Figure 10.2 shows the systematic error e_{sys} as calculated in equation 10.12 for the values considered in this example. There are three curves for different values of w_{fit}. For greater values of w_{fit} the systematic error becomes greater. For $x_{max} = 0$ the systematic error is zero. In this case the samples of our gradient function are symmetric around the maximum sample and every symmetric function fitted through them results in the right estimated value $\hat{x}_{max}$. The systematic bias left and right of the coordinate $x_{max} = 0$ is due to the fact that the gradient of the image signal, which can be approximated by a Gaussian function, is interpolated with a parabolic function. This also explains why the bias is greater at greater values of w_{fit}. The further away the sample positions are from x_{max} the more the samples of the gradient signal differ from a parabolic interpolation. This can be seen in figure 10.1.

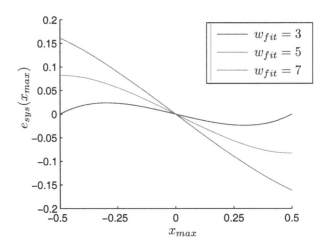

Fig. 10.2. Systematic error $e_{sys}(x_{max})$ for the edge position when a parabolic function is used for interpolation of subpixel values. σ_{all} is set to $\sigma_{all} = 1,4186$.

10.4 Experimental results

In order to verify the calculation from the last section, real world experiments are performed. For the comparison of calculated and measured values, ground truth data is needed. We recorded a scene containing one exact black and white edge. This was achieved by capturing a high resolution LCD showing a black and white edge. The edge lies vertically in the image so that in every row of the image there is one edge location. The camera is slightly rotated around its optical axis to achieve that the filmed edge appears slightly slanted. This way it has the whole range of possible subpixel edge positions from $x_{max} = -0.5$ to $x_{max} = 0.5$. Groundtruth edge positions are determined by fitting a line through all estimated edge positions. To be sure that camera noise is not affecting the analysis an average image was used. 1000 frames of the test image were taken with the camera installed on a tripod. The average value for each pixel position is used for the test image. Figure 10.3 shows a zoomed in view of our test image taken with a Sony DXC-D30WSP 3CCD video camera.

Figure 10.4 shows the measured bias e_m which is given through the difference of ground truth edge positions and estimated edge positions for each line of the test image.

Since the bias e_m repeats periodically for subpixel coordinates of edges from -0.5 to 0.5, it is sufficient to discuss one period of the signal, as can be seen in figure 10.5. Additional to the measured bias e_m, also the estimated bias e_{sys} is shown, for comparison. Measured and estimated values show only small

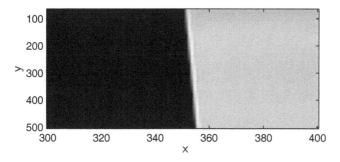

Fig. 10.3. Test image for comparison of measured an calculated values. The edge is slightly slanted.

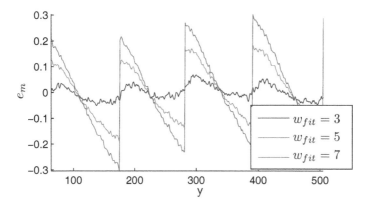

Fig. 10.4. Measured bias e_m for each line of the test image.

differences, that can be explained by small differences in our signal model to that of the real camera.

In [10] a similar edge detector is examined in respect for the uncertainty of edge localisation due to camera noise. A Gaussian distributed localisation error variance of less than $0.002\,\mathrm{pel}^2$ for most edges was found in images with a high PSNR of 42. This corresponds to a standard deviation of $0.044\,\mathrm{pel}$. The maximum bias e_{sys} for a parabolic function through three sample values $w_{fit} = 3$ is about $0.025\,\mathrm{pel}$. This shows that the error introduced by the bias described here has almost the same size as the uncertainty due to camera noise and has to be considered in high level tasks that use edge positions as input data.

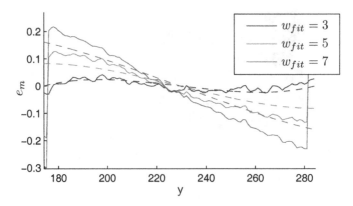

Fig. 10.5. Measured bias e_m for lines 170 to 285 of the test image. The dashed lines show the the estimated bias e_{sys} for comparison.

10.5 Conclusions

This work has shown that edge detectors using parabolic functions for subpixel edge localisation estimate biased values for the edge positions. This is due to the fact that the gradient of the image signal, which can be approximated by a Gaussian function, is interpolated with a parabolic function. This bias is in the same range as errors of the edge positions due to camera noise. Therefore it is advisable to consider this bias in high level tasks that build upon precise edge localisation.

References

1. Canny, J.F.: A computational approach to edge detection. IEEE Transactions on Pattern Analysis and Machine Intelligence **8** (1986) 679–698
2. Kakarala, R., Hero, A.: On achievable accuracy in edge localization. In: IEEE International Conference on Acoustics, Speech, and Signal Processing. Volume 4. (1991) 2545–2548
3. Koplowitz, J., Greco, V.: On the edge location error for local maximum and zero-crossing edge detectors. IEEE Transactions on Pattern Analysis and Machine Intelligence **16** (1994)
4. Tagare, H.D., deFigueiredo, R.J.: On the localization performance measure and optimal edge detection. IEEE Transactions on Pattern Analysis and Machine Intelligence **12** (1990)
5. Verbeek, P.W., van Vliet, L.J.: On the location error of curved edges in low-pass filtered 2-d and 3-d images. IEEE Transactions on Pattern Analysis and Machine Intelligence **16** (1994)
6. Suzuki, K., Horiba, I., Sugie, N.: Neural edge enhancer for supervised edge enhancement from noisy images. IEEE Transactions on Pattern Analysis and Machine Intelligence **25** (2003)

7. Qian, R.J., Huang, T.S.: Optimal edge detection in two-dimensional images. IEEE Transactions on Image Processing **5** (1996)

8. Kisworo, M., Venkatesh, S., West, G.: Modeling edges at subpixel accuracy using the local energy approach. IEEE Transactions on Pattern Analysis and Machine Intelligence **16** (1994)

9. Rockett, P.: The accuracy of sub-pixel localisation in the canny edge detector. Proc. British Machine Vision Conference (1999)

10. Mikulastik, P.: Genauigkeitsanalyse von Kantenpunkten fuer Stereokamerasysteme. In: Niccimon, Das Niedersaechsische Kompetenzzentrum fuer die Mobile Nutzung. Shaker Verlag (2006) 135–143 ISBN 3-8322-5225-2.

11. Mech, R.: Schaetzung der 2D-Form bewegter Objekte in Bildfolgen unter Nutzung von Vorwissen und einer Aperturkompensation. PhD thesis, University of Hannover, Germany (2003) ISBN 3-18-373210-6.

—

Edge Point Linking by Means of Global and Local Schemes

Angel D. Sappa[1] and Boris X. Vintimilla[2]

[1] Computer Vision Center
Edifici O Campus UAB
08193 Bellaterra, Barcelona, Spain
angel.sappa@cvc.uab.es
[2] Vision and Robotics Center
Dept. of Electrical and Computer Science Engineering
Escuela Superior Politecnica del Litoral
Campus Gustavo Galindo, Prosperina, Km 30.5
09015863 Guayaquil, Ecuador
boris.vintimilla@espol.edu.ec

Summary. This chapter presents an efficient technique for linking edge points in order to generate a closed contour representation. The original intensity image, as well as its corresponding edge map, are assumed to be given as input to the algorithm (i.e., an edge map is previously computed by some of the classical edge detector algorithms). The proposed technique consists of two stages. The first stage computes an initial representation by connecting edge points according to a global measure. It relies on the use of graph theory. Spurious edge points are removed by a morphological filter. The second stage finally generates closed contours, linking unconnected edges, by using a local cost function. Experimental results with different intensity images are presented.[3]

11.1 Introduction

Edge detection is the first and most important stage of human visual process as presented in [6]. During last decades several edge point detection algorithms were proposed. In general, these algorithms are based on partial derivatives (first and second derivative operators) of a given image. Unfortunately, computed edge maps usually contain gaps as well as false edge points generated by

---

[3] This work has been partially supported by the Spanish Ministry of Education and Science under project TRA2004-06702/AUT. The first author was supported by The Ramón y Cajal Program. The second author was partially supported by the ESPOL under the VLIR project, Component 8.

noisy data. Moreover, edge points alone generally do not provide meaningful information about the image content, so a high-level structure is required (e.g., to be used by scene understanding algorithms). From a given edge map the most direct high-level representation consists in computing closed contours—linking edge points by proximity, similarity, continuation, closure and symmetry. Something that is very simple and almost a trivial action for the human being, becomes a difficult task when it should be automatically performed.

Different techniques have been presented for linking edge points in order to recover closed contours. According to the way edge map information is used they can be divided into two categories: a) local approaches, which work over every single edge point, and b) global approaches, which work over the whole edge map at the same time. Alternatively, algorithms that combine both approaches or use not only edge map information but also enclosed information (e.g., color) can be found (e.g., [10], [15]). In general, most of the techniques based on local information rely on morphological operators applied over edge points. Former works on edge linking by using morphological operators compute closed boundaries by thinning current edge points [14]. However, common problems of thinning algorithms are that in general they distort the shape of the objects, as well as big gaps can not be properly closed. In order to avoid these problems [12] introduces the use of morphological operators together with chamfer maps. Experimental results with simple synthetic-like images with closely spaced unconnected edges, which do not contain spurious neither noisy edge points, are presented.

A real-time edge-linking algorithm and its VLSI architecture, capable of producing binary edge maps at the video rate, is presented in [7]. It is based on local information and, as stated by the authors, has two major limitations. Firstly, it does not guarantee to produce closed contours, actually in every experimental results presented in that paper there are open contours. Secondly, edge-linking process is sensitive to user defined parameters—threshold values.

In [5], a more elaborated edge linking approach, based only on local information, is proposed. Initially, an iterative edge thinning is applied. Thus, small gaps are filled and endpoints are easily recovered and labelled. Finally, endpoints are linked by minimizing a cost function based on a local knowledge. The proposed cost function takes into account the Euclidean distance between the edge points to be linked (2D distance) and two reward coefficients—a) if the points to be linked are both endpoints; and b) if the direction associated to the points to be linked is opposite. The values of these two reward coefficients are experimentally determined. Since this technique is proposed for linking points, similarly to [7], it does not guarantee to produce closed contours.

Differently to previous approaches, algorithms based on global information need to study the whole edge point distribution at the same time. In general, points are represented as nodes in a graph and the edge linking problem is solved by minimizing some global measure. For instance, [3] presents an edge linking scheme as a graph search problem. A similar scheme was previously introduced in [1]. The methodology consists in associating to every edge point its corresponding gradient—magnitude and direction. Thus, the initial edge map becomes a graph with arcs between nodes ideally unveiling the contour directions. A search algorithm, such as A*, is later on used for finding the best path among the edge points. Although results presented in [3] are promising, the excessive CPU time together with the large number of image dependent parameters, which have to be tuned by the user, discourage its use.

In [11] a fast and free of user-defined parameters technique, which combines global and local information, is presented. It is close related to the previous approaches ([3], [1]), in the sense that graph theory is also used to compute the best set of connections that interrelate edge points. Differently to the previous ones, it is devised to generate closed contours from range image's edge points, instead of classical intensity images. Initially, edge points are linked by minimizing a global cost function. At the same time, noisy data are easily removed by means of an efficient morphological filter. It does not have to go through the whole list of points contained in the input edge map, but only over those points labelled as endpoints—points linked once. In a second stage, closed contours are finally obtained by linking endpoints using a local cost function.

In the current work, we propose to adapt [11] in order to process intensity images. Range image processing techniques can be customized to work with 2D images considering intensity values as depth values. For instance, mesh modelling algorithms, developed for representing 3D images, have been extended to the 2D image field for different applications (e.g., [4], [8], [13]). In the same way, we propose to adapt the contour closure technique presented in [11] in order to handle intensity images.

The remainder of this chapter is organized as follow. Section 11.2 briefly introduces the technique proposed in [11] together with the required changes to face up intensity images. Experimental results with several images are presented in Section 11.3. Finally, conclusions and further improvements are given in Section 11.4.

11.2 Proposed Technique

Let I be a 2D array representing an intensity image with R rows and C columns, where each array element $I(r,c)$ is a value defined as $0 \leq I(r,c) \leq 255$. In order to have a direct application of the approach proposed in [11], intensity values are considered as depth values; so every pixel in $I(r,c)$ becomes a point in 3D space: $(x,y,z) = (r,c,I(r,c))$. Let E be the corresponding edge map computed by an edge point detector algorithm. Each element of $E(r,c)$ is a boolean indicating whether the corresponding image pixel is an edge point or not. In the current implementation edge maps were computed by using Canny edge detector [2]. Additionally, edge points uniformly distributed through the first and last rows and columns were added. Added edge points are useful for detecting a region boundary when it touch an image's border; actually, the

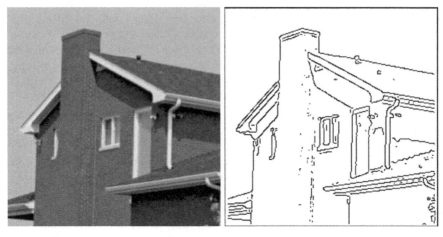

Fig. 11.1. (*left*) Input intensity image, I. (*right*) Input edge map, E, computed by Canny.

idea of imposing edge points through the image border has already been used in [9].

Assuming both arrays, I and E (see Fig. 11.1), are given as inputs the proposed technique consists of two stages. The first stage links edge points by minimizing a global measure. Computed connections are later on filtered by means of a morphological operator. The second stage works locally and is only focussed on points labelled as endpoints. Both stages are further described below.

11.2.1 Global Scheme: Graph Based Linking

At this stage a single polyline that links all the input edge points, by minimizing the sum of linking costs, is computed. On the contrary to [5], where a linking cost considering the Euclidean distance in the edge map is used (distance in a 2D space), we propose to use also the Euclidean distance but in the 3D space. Neighbor points in the edge map could belong to different regions in the intensity image. In other words, using only point positions in the edge map could drive to wrong results. Therefore, linking cost between edge points $(E_{(i,j)}, E_{(u,v)})$ is defined as:

$$LC_{(i,j),(u,v)} = \|(i, j, I(i, j)) - (u, v, I(u, v))\| \tag{11.1}$$

In order to speed up further processing, a partially connected graph Γ is computed, instead of working with a fully connected one. Since this partially connected graph should link nearest neighbor edge points, a 2D Delaunay triangulation of the edge map's points is computed. Additionally, every edge is associated with a cost value computed as indicated above, $LC_{(i,j),(u,v)}$.

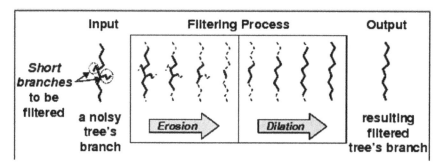

Fig. 11.2. Filtering process: *opening algorithm.*

Finally, the shortest path in Γ that links all the edge points is extracted by computing the Minimum Spanning Tree (MST) of Γ. The MST of Γ is the acyclic subgraph of Γ that contains all the nodes and such that the sum of the costs associated with its edges is minimum. Notice that the MST of the Delaunay triangulated input edge points gives the same result than if it were computed over a fully connected graph of those points.

Fig. 11.3(*top*) shows the triangular mesh and its corresponding MST, computed from Fig. 11.1; input edge map, Fig. 11.1(*right*), contains edge points computed by the edge detector [2], as well as edge points added over the first and last rows and columns. As can be appreciated in Fig. 11.3(*top−right*), the resulting MST contains short branches—branches defined by a few edges—, connected with the main path. They belong to information redundancy and noisy data. So, before finishing this global approach stage, and taking advantage of edge point connections structured as a single polyline, a morphological filter is applied. The filter is a kind of *opening algorithm* and consists in performing iteratively erosions followed by the corresponding dilations; the latter applied as many times as the erosion. In brief, the opening algorithm considers segments of the polyline as basic processing elements (like pixels in an intensity image). From the polyline computed by the MST, those segments linked from only one of their defining points—referred as *end segments*—are removed during the erosion stage. After ending the erosion process, dilations are carried out over end segments left. Fig. 11.2 shows and illustration of this filtering process; in this case it consists of four dilations applied after four erosions. More details about the filtering process can be found in [11].

11.2.2 Local Scheme: Cost Based Closure

The outcome of the previous stage is a single polyline going through almost all edge points (some edge points were removed during the last filtering stage). Fig. 11.3(*bottom−left*) presents the result obtained after filtering the MST of Fig. 11.3(*top−right*). Notice that although this polyline connects edge points it does not define closed contours—recall that the MST is an acyclic subgraph

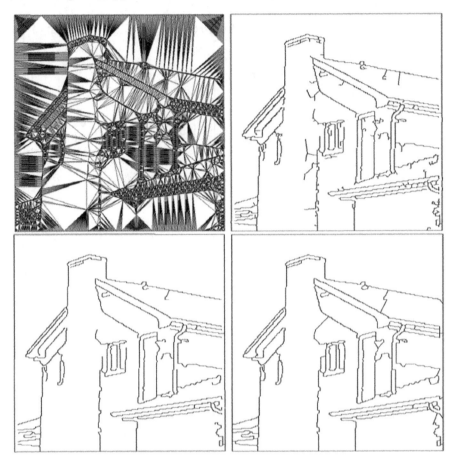

Fig. 11.3. (*top* − *left*) Triangular mesh of the edge points presented in Fig. 11.1(*right*). (*top* − *right*) Minimum spanning tree. (*bottom* − *left*) Filtered MST—opening algorithm. (*bottom* − *right*) Final linked edge point representation.

so that it does not contain any closed contours. Therefore, the objective at this last stage focuses on closing open contours.

Open contours are characterized by edge points linked once—endpoints. Since the previous filtering stage was carried out over end segments, endpoints are easily identified; there is no need to go through the whole list of edge points to find those only linked once. For every endpoint a list of candidate points from the edge map E is extracted. Finally, the point with a minimum closure cost is chosen to close the given endpoint. These stages are detailed below.

Given an end point $E(i,j)$, its set of candidate edge points is selected by means of an iterative process over a dynamic window, DW, centered at that point—$DW_{(i\pm m, j\pm n)}$, where $m = \{1, \ldots, t\}$; $n = \{1, \ldots, t\}$; $t = s + \tau$; and $\{(s < m < t) \vee (s < n < t)\}$. During the first iteration s is set to zero.

Then after each iteration it is increased by τ. The threshold τ depends on the density of edge points in the given edge map; in the current implementation τ was set to four.

After extracting the set of candidate points from the current iteration, a closure cost, CC, is computed. It represents the cost of connecting each one of those candidates with the given endpoint $E(i,j)$. It is computed according to the following expression:

$$CC_{(i,j),(u,v)} = \frac{LC_{(i,j),(u,v)}}{PathLength_{(i,j),(u,v)}} \qquad (11.2)$$

$LC_{(i,j),(u,v)}$ is the linking cost defined in (11.1), which represents the 3D distance between the points to be linked; while $PathLength_{(i,j),(u,v)}$ measure the length of the path—number of edges—linking those two points. In case of no candidate points were extracted from the current window or $PathLength_{(i,j),(u,v)}$ values from those candidates to the given endpoint were equal or smaller than t, the size of DW is increased by τ, so that s and t, and the process starts again by extracting a new set of candidate points. The new set of candidate points does not contain those previously studied due to the fact that the new window is only defined by the outside band. Otherwise, the point with lowest closure cost is chosen to be linked with the endpoint $E(i,j)$.

11.3 Experimental Results

The proposed technique has been tested with different intensity images. As mentioned above, in all the cases edge maps were computed by using Canny edge detector [2]. Additionally, a set of edge points uniformly distributed over the image border (first and last rows and columns) was added. The CPU time to compute the different stages have been measured on a 1.86 GHz Pentium M PC with a non-optimized C code.

The illustrations used through the chapter correspond to an intensity image of 256×256 pixels (Fig. 11.1(*left*)) and an edge map defined by 4784 points (Fig. 11.1(*right*)); its MST contains 4783 edges and was computed in 0.56 sec (Fig. 11.3(*top − right*)). The opening algorithm filters 378 edges from the computed MST giving rise to a representation with 4405 edges in 0.03 sec (Fig. 11.3(*bottom − left*)). The 378 removed edges correspond to those ones linked with noisy data or redundant edge points. Finally, 52 open contours are closed in 0.05 sec. This final representation contains 4457 edges, Fig. 11.3(*bottom − right*)

Other images were processed with the proposed approach. Fig. 11.4(*top − left*) presents the input edge map, 21393 points, corresponding to an image of 512×512 pixels. Intermediate results, such as Delaunay triangulation of input edge points and its MST, are also presented in Fig. 11.4. The MST is defined by 21392 edges. The final closed contour representation is presented

Fig. 11.4. (*top − left*) Input edge points, 21393 points. (*top − right*) Triangular mesh. (*bottom − left*) Minimum spanning tree, 21392 edges. (*bottom − right*) Final closed contour representation (after filtering the MST and closing open boundaries), 18517 edges.

in Fig. 11.4(*bottom − right*); it contains 18517 edges and was computed in 28.4 sec. Have a look at those gaps on the shoulder and top of the hat that are successfully closed in the final representation.

Finally, Fig. 11.5(*top*) shows edge maps defined by 6387 and 34827 points respectively; the corresponding intensity images are defined by 256×256 pixels (girl) and 512×512 pixels (car). The results from the global approach stage are presented in Fig. 11.5(*middle*)—filtered MST. Final results are given in Fig. 11.5(*bottom*); they are defined by 5052 and 27257 edges respectively. Information regarding CPU time for the different examples are presented in Tab. 11.1. As can be appreciated in all the examples about 85% of the time

Fig. 11.5. (*top*) Input edge points (6387 and 34827 points). (*middle*) Filtered MST (4964 and 26689 edges). (*bottom*) Final closed contour representation (5052 and 27257 edges).

Table 11.1. CPU time (sec)

	Global Scheme		Local Scheme	Total
	Triangular Mesh and MST Generation	Filtering	Contour Closure	Time
House	0.56	0.03	0.05	0.65
Lenna	23.9	0.76	3.74	28.41
Car	80.43	2.35	11.67	94.46
Girl	0.98	0.07	0.13	1.19

is spent by the triangular mesh and MST generation. Since a non-optimized C code is used, it is supposed that there is a room for improvement.

11.4 Conclusions and Further Improvements

This chapter presents the use of global and local schemes for computing closed contours from edge points of intensity images. The global stage is based on graph theory while the local one relies on values computed by a local cost function. Noisy and redundant edge points are removed by means of an efficient morphological operator. Although this approach has been initially proposed to handle range images, experimental results proved that it is also useful for processing intensity images.

Further work will be focused on improving MST generation, for instance by generating it at the same time that the triangular mesh. Additionally, the development of new linking cost and closure cost functions, specifically designed for handling intensity images, will be considered. It is supposed that cost functions that take into account information such as intensity of pixels crossed by the graph edges could improve the final results. Finally, comparisons with other approaches will be done.

References

1. D. Ballard and C. Brown. *Computer Vision.* Prentice-Hall, Inc., 1982.
2. J. Canny. Computational approach to edge detection. *IEEE Trans. Pattern Analysis and Machine Intelligence*, 8(6):679–698, 1986.
3. A. Farag and E. Delp. Edge linking by sequential search. *Pattern Recognition*, 28(5):611–633, May 1995.
4. M.A. Garcia, B. Vintimilla, and A. Sappa. Approximation and processing of intensity images with discontinuity-preserving adaptive triangular meshes. In *Proc. ECCV 2000, D. Vernon, Ed. New York: Springer, 2000, vol. 1842, LNCS, Dublin, Ireland*, pages 844–855, June/July 2000.
5. O. Ghita and P. Whelan. Computational approach for edge linking. *Journal of Electronic Imaging*, 11(4):479–485, October 2002.

6. W.E. Grimson. *From Images to Surfaces*. Cambridge, MA: MIT Press, 1981.
7. A. Hajjar and T. Chen. A VLSI architecture for real-time edge linking. *IEEE Trans. on Pattern Analysis and Machine Intelligence*, 21(1):89–94, January 1999.
8. L. Hermes and J. Buhmann. A minimum entropy approach to adaptive image polygonization. *IEEE Trans. on Image Processing*, 12(10):1243–1258, October 2003.
9. C. Kim and J. Hwang. Fast and automatic video object segmentation and tracking for content-based applications. *IEEE Trans. on Circuits and Systems for Video Technology*, 12(2):122–1129, February 2002.
10. E. Saber and A. Tekalp. Integration of color, edge, shape, and texture features for automatic region-based image annotation and retrieval. *Journal of Electronic Imaging*, 7(3):684–700, July 1998.
11. A.D. Sappa. Unsupervised contour closure algorithm for range image edge-based segmentation. *IEEE Trans. on Image Processing*, 15(2):377–384, February 2006.
12. W. Snyder, R. Groshong, M. Hsiao, K. Boone, and T. Hudacko. Closing gaps in edges and surfaces. *Image and Vision Computing*, 10(8):523–531, October 1992.
13. Y. Yang, M. Wernick, and J. Brankov. A fast approach for accurate content-adaptive mesh generation. *IEEE Trans. on Image Processing*, 12(8):866–881, August 2003.
14. T. Zhang and C. Suen. A fast parallel algorithm for thinning digital patterns. *Communications of the ACM*, 27(3):236–239, March 1984.
15. S. Zhu and A. Yuille. Region competition: Unifying snakes, region growing, and bayes/mdl for multiband image segmentation. *IEEE Trans. Pattern Analysis and Machine Intelligence*, 18(9):884–900, 1996.

12

An Enhanced Detector of Blurred and Noisy Edges

M. Sarifuddin[1], Rokia Missaoui[1], Michel Paindavoine[2] and Jean Vaillancourt[1]

[1] Département d'informatique et d'ingénierie, Université du Québec en Outaouais, C.P. 1250, Succ. B, Gatineau (Qc), Canada, J8X 3X7
e-mail : {rokia.missaoui, m.sarifuddin, jean.vaillancourt}@uqo.ca,
[2] Laboratoire LE2I - UMR-CNRS, Université de Bourgogne, Aile des Sciences de l'Ingénieur, 21078 Dijon cedex - France, e-mail : paindav@u-bourgogne.fr

Summary. Detecting edges in digital images is a tricky operation in image processing since images may contain areas with different degrees of noise, blurring and sharpness. Such operation represents an important step of the whole process of similarity shape analysis and retrieval.

This chapter presents two smoothing and detection filters which are based on a model of blurred contours and well adapted to the detection of blurred and/or noisy edges. These filters can be implemented in a third-order recursive form and offer advantages in the analysis of different edge types (sharp, noisy and blurred). Experimental analysis shows that these filters give definitely better edge detection and localization than some existing filters.

Key words: Blurred edge model, smoothing filter, edge detection filter.

12.1 Introduction

Content-based image retrieval (CBIR) makes use of lower-level features like color, texture, spatial layout and shape, and even higher-level (semantic) features like annotations and user interactions to retrieve images according to different search paradigms. Research studies on edge detection frequently assume that object shapes are already stored in the database (e.g., Surrey database) rather than computed within the retrieval process. However, the identification of edges inside digital images is a non-trivial operation in image processing. It is more complicated when images contain not only sharp, but also blurred and noisy regions. There are many algorithms for shape description such as Fourier descriptor [1], moment invariants, B-Spline [2], active shape models, wavelet descriptor [3] and statistical shape Analysis [4].

The objective of this chapter is to propose two filters: one for smoothing and the other one for edge detection. The first filter minimizes or eliminates noise whereas the second one allows edge detection in digital images. The detection filter should take into account the fact that image smoothing will result in blurred edges. For this reason, the development of the two filters is based on a blurred edge model.

12.2 Edge Detectors

Sharpness of object edges during an image capture depends on the diaphragm opening, shutter speed of the camera, and the light effect in the scene. The moving objects in the scene will be blurred and the rest of the image will be sharp if the shutter speed is slow. However, the sharpness of the scene can involve noise which is more visible in dark image areas. Therefore, we can state that, by nature, a real image taken by a photo/video camera may contain sharp, blurred, and noisy areas. Consequently, image analysis techniques have to take into account these characteristics. An edge in a given image is a set of pixels corresponding to abrupt changes of intensity value. Hence, edge detection is the process that identifies the pixels belonging to an edge. It can be determined based on the gradient or Laplacian. In case of the gradient, the local maxima correspond to the edge, while in the case of Laplacian, the zero-crossing represents the edge.

The performance of a given detector is closely related to the computation time and the detection efficiency. A detector efficiency is often based on three Canny's criteria [5]:

- *good detection*: all edges have to be detected (without losing certain pixels on the edge to be found),
- *good localization*: the edges have to be located in their ideal positions,
- *low multiplicity of the response*: the detector does not give multiple responses or false edges.

There are many factors which can influence the performance of an edge detection process in real images. They include noise, blurring and the interference between adjacent edges. Moreover, an edge can be located in a sharp area or submerged into a noisy or blurred area. To get an appropriate detection, it is necessary to use a detector which takes into account the particular features of the area. The elimination of the harmful effects of these factors (e.g., noise) has lead to the development of many algorithms for edge detection such as first-order operators (Roberts, Sobel and Prewitt operators), second-order operators (Laplacian, LOG and DOG operators), multi-scale algorithms, and filtering algorithms.

The first-order operators act as a high frequency filter, which cause high noise sensitivity of the corresponding algorithms. Different existing variations - the Roberts operator[6] is one of them - bring some improvement to the discrete estimation of the gradient (derivative) and take into account the edge orientation. Prewitt operator[8] and Sobel operator[7] incorporate explicitly in their filters a smoothing operation in order to attenuate moderately the noise influence. In algorithms based on the Laplacian or second order operators, the edge pixels are located in the zero-crossing positions of the image Laplacian (approximation of the second-order derivative of the image). This approach has the advantage to provide thin and closed edges but has a higher sensitivity to noise compared to the gradient approach. Different methods for the estimation of the Laplacian generated a variety of algorithms. In the case of the Marr-Hildreth operator[9], the Gaussian smoothing is incorporated in the form of a LOG filter (Laplacian of Gaussian).

Multi-scale algorithms decompose an image into spatio-temporal subbands by high-pass and low-pass filters and then make the fusion of the detection results. The method proposed in [10] uses a multi-scale wavelet decomposition. The filters of Canny [5], Deriche [11], Bourennane [12], Laggoune [13] and Demigny [14] can be included into the group of optimized filters. Each one of these detectors has its own properties and peculiarities. Canny [5], Deriche [11] and Demigny [14] have proposed edge detectors based on a step edge model whereas the Bourennane filter [12] and Laggoune filter [13] use the edge model of ramp and crest-line, respectively.

12.3 Filter Definition

In this section we present new filters, which are based on a blurred edge model. The first part of this section will describe the edge models whereas the second one will present the proposed filters.

12.3.1 Model of Blurred Edges

Edge detection in an image is a very important feature in the process of object recognition and content-based image retrieval. Development of an edge detector is often based on a specific property of the image. For example, when considering the previously mentioned authors [5, 11, 14, 12, 13], each of these researchers proposes a detector that is well adapted to sharp and noisy images by a tuning of one or several parameters.

Figures 12.1-a to 12.1-c represent the models of step, crest-line and impulse edges respectively, while figures 12.1-d to 12.1-f show the blurred version of the previous edge models.

In our case, we select Figure 12.1-e as a blurred edge model. According to its plot, the model of a blurred crest-line edge can be mathematically represented by a function $C(x)$ (Equation 12.1) with two normalizing constants K and λ, the blurring parameter β and the scale parameter α related to noise. The value of the blurring parameter β varies between 0 and 1. When β is close to zero, $C(x)$ corresponds to a non-blurred crest-line edge model, whereas when β is close to 1, $C(x)$ represents a blurred crest-line edge model.

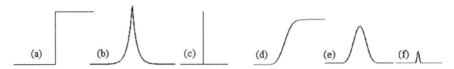

Fig. 12.1. (a) step edge, (b) crest-line edge and (c) impulse edge. (d) Blurred step edge (ramp edge), (e) blurred crest-line edge and (f) blurred impulse edge.

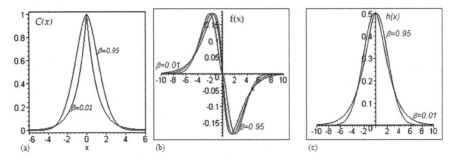

Fig. 12.2. (a) Model of blurred crest-line edges, (b) Plot of the detection filter $f(x)$ and (c) Plot of the smoothing filter $h(x)$.

In addition, when α is zero, $C(x)$ corresponds to an impulse edge. Figure 12.2-a shows the curve of this function for $\alpha=1$ and two distinct values of β.

$$C(x) = K.e^{-\alpha|x|}.(\lambda + \frac{\sin(\alpha\beta|x|)}{\beta} - \cos(\alpha\beta|x|)) \tag{12.1}$$

12.3.2 Edge Detection Filters

Let us suppose that $C(x)$ corresponds to an edge model in an image (model of the edge intensity variation of crest-line type along the axis x) described by Equation 12.1. The edges in the image can be computed by a second derivative (Laplacian) or by a first derivative (gradient) of the function $C(x)$ according to the following formula:

$$f(x) = \frac{\partial C(x)}{\partial x} = -K.\alpha.e^{-\alpha|x|}.(\lambda - 2\cos(\alpha\beta|x|) + \frac{(1-\beta^2)}{\beta}\sin(\alpha\beta|x|)) \tag{12.2}$$

We require that $f(x)$ becomes a band-pass filter. Its impulse response has to be an odd function, i.e. $\int f(x)dx = 0$, $f(0) = 0$, and $f(x) = -f(-x)$. In order to make it stable, this function should be convergent, i.e. $f(-\infty) = 0$ et $f(\infty) = 0$. Therefore, the value of λ can be derived as follows: $f(0) = 0 = -K\alpha(\lambda - 2) \Rightarrow \lambda = 2$. In order to guarantee that the function is odd, we can generalize Equation 12.2 as follows (see Equation 12.3), with $K_1 = -2K\alpha$

which corresponds to the new normalization constant and $sgn(-x)$ returns the negative sign when x is positive and the positive sign when x is negative. The plot of this filter is illustrated in Figure 12.2-b for $\alpha = 0.75$ with different values of β (β=0.01, β=0.65 and β=0.95). We notice that the extrema of the curve move away from the axis y as much as the value of β increases and approaches to one. This indicates that the filter $f(x)$ can detect blurred or sharp edges.

$$f(x) = sgn(-x).K_1.e^{-\alpha|x|}.(1 - \cos(\alpha\beta|x|)) + \frac{(1 - \beta^2)}{2\beta} \sin(\alpha\beta|x|)) \quad (12.3)$$

The discrete expression for $f(x)$ filter can be obtained using the Z-transform, which is the sum of the causal and anti-causal parts. Each one of the two parts is implemented in a third-order recursive form. The impulse response $y(m)$ of digital signal $S(m)$ corresponding to the $f(x)$ digital filter can be written as:

$$y(m) = y^+(m) - y^-(m) \quad (12.4)$$

where $y^+(m)$ determines a left-to-right recursion (causal parts) and $y^-(m)$ determines a right-to-left recursion (anti-causal parts) given by:

$$y^+(m) = K_1(c_0.s(m - 1) + c_1.s(m - 2)) - b_1.y^+(m - 1) - b_2.y^+(m - 2) - b_3.y^+(m - 3)$$

$$y^-(m) = K_1(c_0.s(m + 1) + c_1.s(m + 2)) - b_1.y^-(m + 1) - b_2.y^-(m + 2) - b_3.y^-(m + 3)$$

where the coefficients c_i and b_i are given by:
$c_0 = e^{-\alpha}(1 - \cos(\alpha\beta) + \frac{(1-\beta^2)}{2\beta} \sin(\alpha\beta))$, $c_1 = e^{-2\alpha}(1 - \cos(\alpha\beta) - \frac{(1-\beta^2)}{2\beta} \sin(\alpha\beta))$,
$b_1 = -e^{-\alpha}(1 + 2\cos(\alpha\beta))$, $b_2 = e^{-2\alpha}(1 + 2\cos(\alpha\beta))$, $b_3 = -e^{-3\alpha}$ and
$K_1 = \frac{1+b_1+b_2+b_3}{c_0+c_1}$.

12.3.3 Smoothing Filters

In the case of noisy image analysis, it is first necessary to attenuate or smooth the noise and then to detect edges. The smoothing filter $h(x)$ can be determined as follows:

$$Smoothed\ signal: \qquad S'(x) = h(x) * S(x); \quad (12.5)$$

The signal edges correspond to the derivative of the smoothed signal or can be calculated by the convolution of the detection filter and the signal as follows:

$$Edge\ of\ signal: \qquad f(x) * S(x) \approx \frac{\partial(S'(x))}{\partial x} = \frac{\partial h(x)}{\partial x} * S(x); \quad (12.6)$$

where $S(x)$ is 1-D signal and $f(x)$ is equivalent to the first derivative of a smoothing filter. So, the smoothing filter using K_2, which is a normalization constant, can be calculated as follows:

$$h(x) = \int f(x)dx = K_2.e^{-\alpha|x|}(1 - \frac{1}{2}\cos(\alpha\beta|x|) + \frac{1}{2\beta}\sin(\alpha\beta|x|)) \quad (12.7)$$

The plot of this filter is illustrated in Figure 12.2-c for $\alpha = 0.75$ with different values of β (β=0.01 and β=0.95). The discrete expression for $h(x)$ filter can be obtained using the Z-transform, which is the sum of the causal and anti-causal parts. Each one of the two parts is implemented in a third-order recursive form. The impulse response $y(n)$ of digital signal $S(n)$ corresponding to the $h(x)$ digital filter can be written as:

$$y(n) = y^+(n) + y^-(n) \quad (12.8)$$

where $y^+(n)$ determines a left-to-right recursion (causal parts) and $y^-(n)$ determines a right-to-left recursion (anti-causal parts) given by:

$y^+(n) = K_2(a_0.s(n) + a_1.s(n-1) + a_2.s(n-2)) - b_1.y^+(n-1) - b_2.y^+(n-2) - b_3.y^+(n-3)$

$y^-(n) = K_2(a_3.s(n+1) + a_4.s(n+2) + a_5.s(n+3)) - b_1.y^-(n+1) - b_2.y^-(n+2) - b_3.y^-(n+3)$

where the coefficients a_i take the following values:
$a_0 = 0.5$, $a_1 = e^{-\alpha}(0.5 - 1.5\cos(\alpha\beta) + \frac{0.5}{\beta}\sin(\alpha\beta))$, $a_2 = e^{-2\alpha}(1 - 0.5\cos(\alpha\beta) - \frac{0.5}{\beta}\sin(\alpha\beta))$, $a_3 = e^{-\alpha}(1 - 0.5\cos(\alpha\beta) + \frac{0.5}{\beta}\sin(\alpha\beta))$, $a_4 = e^{-2\alpha}(0.5 - 1.5\cos(\alpha\beta) - \frac{0.5}{\beta}\sin(\alpha\beta))$, $a_5 = 0.5e^{-3\alpha}$ and $K_2 = \frac{1+b_1+b_2+b_3}{a_0+a_1+a_2+a_3+a_4+a_5}$.

12.3.4 Algorithm for 2-D Edge Detection

The 2-D representation of the edge detection filter can be obtained using a convolution between the 1-D smoothing function with 1-D detection function imposing the separability condition with respect to x and y axes. The two filters $f^x(x,y) = f(x) * h(y)$ and $f^y(x,y) = h(x) * f(y)$ (see Figures 12.3-a and 12.3-b) are obtained as follows: in order to have a cost-effective recursive 2-D implementation, the separability of the filters for image analysis in two directions x (or m) and y (or n) is considered [11]. It means that the smoothing in x is followed by the detection in y (horizontal edges) and the detection in x is followed by the smoothing in y (vertical edges). The algorithm given below represents the 2-D edge detection by the convolution of the image $I(m,n)$ by the proposed filters. $F(n)$ and $F(m)$ are the digital filters of $f(x)$ for the axes n and m respectively, and $H(n)$ and $H(m)$ are the digital filters for $h(x)$.

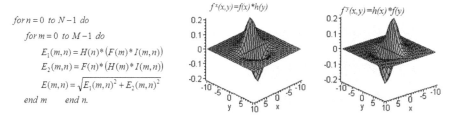

$$for \ n = 0 \ to \ N-1 \ do$$
$$for \ m = 0 \ to \ M-1 \ do$$
$$E_1(m,n) = H(n)*(F(m)*I(m,n))$$
$$E_2(m,n) = F(n)*(H(m)*I(m,n))$$
$$E(m,n) = \sqrt{E_1(m,n)^2 + E_2(m,n)^2}$$
$$end \ m \qquad end \ n.$$

Fig. 12.3. 2-D edge detection algorithm and plot of the 2-D edge detection filters.

12.4 Experimental Results

In this section we present the results of edge detection obtained by the proposed $f(x)$ and $h(x)$ filters and the Deriche filters. Figure 12.4-a shows synthetic images containing a step edge and crest-line edge smoothed by the Gaussian with the standard deviation $\sigma = 10$ and then altered by white Gaussian noise (SNR=2). Figures 12.4-c and 12.4-d show the edges of the blurred and noisy image (Figure 12.4-a) obtained by the application of the Deriche filters for $\alpha = 0.35$ and our filters for $\alpha = 0.35$ and $\beta = 0.25$, respectively. Figure 12.5-a shows synthetic images, containing six different objects, smoothed by the Gaussian with the standard deviation $\sigma = 4$ and then altered by white Gaussian noise (SNR=4). Figures 12.5-c and 12.5-d represent the edges of the blurred and noisy image (Figure 12.5-a) obtained by the application of the Deriche filters for $\alpha = 0.85$ and our filters for $\alpha = 0.85$ and $\beta = 0.25$, respectively. We can conclude that the filters proposed in this chapter give better results compared to the Deriche filters. For example, for the blurred and noisy image (Figure 12.4), with $\alpha = 0.35$, the Deriche detector gives a multiple edge response while our detector gives a single edge response.

We have used the proposed filters to estimate tree growth (tree age) by identifying rings in wood samples. Figure 12.6 shows a slice of wood and the rings detected by Deriche and our filters.

12.4.1 local edge detection

It is difficult to detect effectively edges if a single operator or a filter is applied to the whole image which may contain regions with different properties relatively to noise, blurring and sharpness. In order to improve the detection efficiency, we propose to segment the image into homogeneous regions according to criteria of noise, blurring or sharpness presence. The following labels are assigned to each segmented region: blurred region, noisy region, sharp-uniform or quasi-uniform region and sharp region with detail levels. Then, an appropriate edge detector is applied to each image region based on its peculiarities.

134

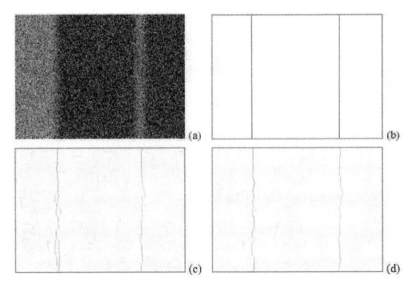

(a)

(b)

(c)

(d)

Fig. 12.4. (a) A synthetic image, with step and crest-line edges, blurred by the Gaussian and altered by white Gaussian noise. (b) The ideal edges of image (a), (c) Edges obtained by the Deriche filters for $\alpha = 0.35$ and (d) Edges obtained by our filters for $\alpha = 0.35$ and $\beta = 0.25$.

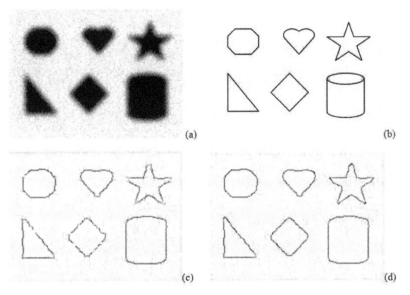

(a)

(b)

(c)

(d)

Fig. 12.5. (a) A synthetic image containing six different objects, blurred by the Gaussian ($\sigma = 4$) and altered by white Gaussian noise (SNR=4).(b) The ideal edges of image (a),(c) Image edge from Figure 5-a obtained by the Deriche filters for $\alpha = 0.85$ and (d) Edges obtained by our filters for $\alpha = 0.85$ and $\beta = 0.25$.

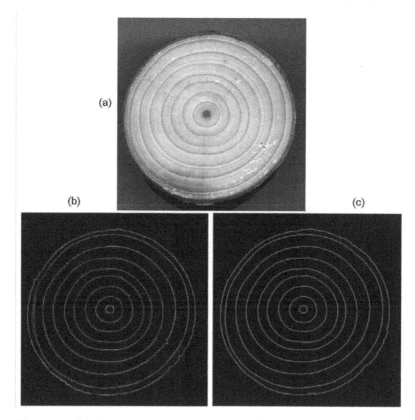

Fig. 12.6. (a) Original image of a wood slice.(b) Ring shapes obtained by Deriche filters for $\alpha = 1.125$. (c) Ring shapes obtained by our filters for $\alpha = 1.125$ and $\beta = 0.75$.

We consider a region as being sharp uniform (smooth surface) if the mean value of the pixel intensity variance is very small or almost zero (few edges). A region is called sharp with detail levels if edges are present and the significant variation of intensity gives a visual impression of a sharp separating line between two adjacent plateaus in the image. The detail levels are measured using the histogram and the entropy of the image region. In the presence of blurred edges, the edges are available but intensity variation is weak (gradual transition) at the border of two adjacent plateaus in the image. A region is noisy if there is a presence and/or absence of edges whose pixel intensity (signal amplitude and frequency) varies randomly over the entire region. Figure 12.7 illustrates the Lena image containing edges in sharp, noisy (added noise), and blurred regions. In such a case of heterogeneity, it is important to apply an adaptive edge detection technique to each one of the regions by selecting the most appropriate detector to the properties of the region. In order to identify the property of each image region we adopt the algorithm for noise and blurring estimation, which is defined in [15].

Blurred area

Noisy area

Area with many details

Uniform area

Fig. 12.7. Image Lena containing the properties of blurring, noise, and detail levels

Figure 12.8 shows the application of our filter using the parameters fixed for the whole image. In the first case, all the edges in sharp areas are well detected maintaining their good localization. However, in the second case, one can notice that the blurred and noisy edges are well detected but the sharp edges are displaced (see edges at the hair level). Figures 12.9-a and 12.9-b show in a qualitative manner the result of area segmentation (sharp, blurred, and noisy) followed by the utilization of our filter with parameters α and β, which are adjusted dynamically relatively to the area property. This result shows clearly the good quality of detection while maintaining the good edge localization in various homogeneous image areas.

12.5 Filter Performance

In this section we analyze in a quantitative manner the performance of the filters. First, we handle the theoretical performance of our detection filter, and then, we discuss the experimental performance.

As mentioned in Section 12.2, the performance of the filter can be determined based on three criteria: good detection, good localization, and low multiplicity of the response (uniqueness). We apply these criteria to the evaluation of our detection filter. According to Canny criteria, an optimal edge

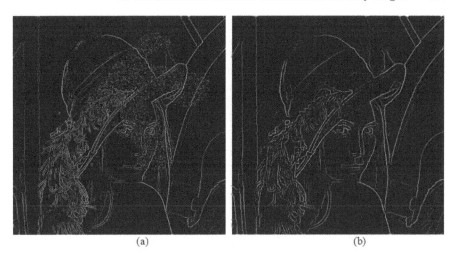

(a)   (b)

Fig. 12.8. Result of edge detection applying our filter with the parameters fixed for the whole image. (a) with parameters $\alpha = 1.5$ and $\beta = 0.5$ and (b) with $\alpha = 0.75$ and $\beta = 0.5$.

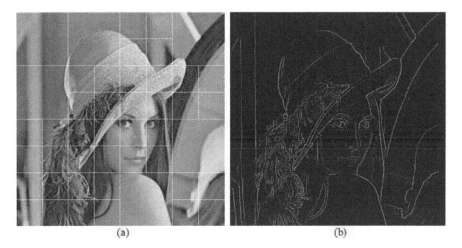

(a)   (b)

Fig. 12.9. (a) Segmentation of image Lena into sharp, blurred and noisy (homogeneous regions). (b) Edge detection results based on the homogeneous regions after the segmentation.

detection filter provides the maximum of products $\Sigma.L$ and $\Sigma.L.U$, where Σ is the signal to noise ratio SNR, L is the localization index (Pratt index) and U is multiplicity of the response. Deriche filter has a performance $\Sigma.L = 1.12$. Figures 12.10-a and 12.10-b show the influence of the parameters α and β of our filter $f(x)$ on the products $\Sigma.L$ and $\Sigma.L.U$, respectively. The curve in Figure 12.10-a shows that the filter $f(x)$ is optimal for $\beta = 0.87, \forall \alpha$, (product $\Sigma.L = 2.52$) and the curve in Figure 12.10-b (product $\Sigma.L.U$) shows that the filter $f(x)$ is optimal for $\beta = 0.77, \forall \alpha$.

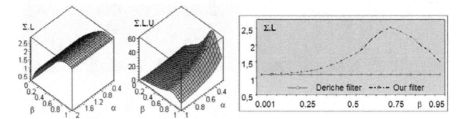

Fig. 12.10. (a) and (b) Theoretical curves of the products $\Sigma.L$ and $\Sigma.L.U$ resp. for our filter and (c) Experimental comparison of the product $\Sigma.L$ between the Deriche filter and our filter.

The following equations provide the definition of SNR, L and U respectively.

$$SNR = 10.log(\frac{N_b \sum_{i,j} P_c(i,j)^2}{N_c \sum_{i,j} P_b(i,j)^2}) \tag{12.9}$$

$$L = \frac{1}{N_{cm}} \sum_{l=1}^{N_c} \frac{1}{1 + d_l^2} \tag{12.10}$$

$$U = \frac{N_I}{N_{cm}} \tag{12.11}$$

$N_{cm} = max(N_c, N_I)$, N_c and N_I correspond to the number of pixels belonging to the detected edges and ideal edges, respectively. N_b is the number of noisy pixels, P_c and P_b represent the edge pixel intensity and the noise pixel intensity, respectively. d_l is the distance between the pixel of an ideal edge and the pixel of a detected edge at the position l. Figure 12.10-c shows the experimental performance of the Deriche detection filter and our filter in terms of good detection and good localization. This curve shows that our filter has a better performance than the Deriche filter. It gives the maximum of the product $\Sigma.L$ for the value of β near 0.75 and any value of α.

12.6 Conclusion

We have described two new filters for edge smoothing and detection that depend on two parameters. The parameter β tunes the blurring influence on the edges, whereas the parameter α is important for noise elimination. Our filters clearly outperform the Deriche filters for $\beta > 0$, $\forall \alpha$, and are optimal for $0.7 < \beta < 0.9$ and any value of α. Theoretical and experimental results show that our smoothing and detection filters are better adapted to sharp, blurred, noisy or blurred and noisy images.

References

1. C. T. Zahn and R. Z. Roskies, : Fourier descriptors for plane closed curves. IEEE Trans. on Comput. 3. C-21 (1972) 269-281
2. F. S. Cohen and Z. Huang, Z. Yang,: Invariant matching and identification of curves using B-Splines curve representation. IEEE Trans. On Image Processing. Vol. 4. No. 1 (1995) 1–10
3. G.C.H. Chuang and C.C.J. Kuo, : Wavelet Descriptor of Planar Curves: Theory and Applications. IEEE Tans. on Image Processing. Vol. 5, No. 1, (1996) 56–70
4. A. Srivastava and S. H. Joshi and W. Moi and X. Liu, : Statistical Shape Analysis : Clustering, Learning and Testing. IEEE Trans. On PAMI. Vol. 27, No. 4, (2005)
5. J. Canny : A Computational approach to edge detection. IEEE Trans. On PAMI. Vol. 8, (1986) 679–697
6. L.G. Roberts, : Machine perception of three dimensional solids. In J.T. Tippet, editor, Optical and Electro-optical Information Processing, pages 159–197. MIT Press, 1965.
7. I. Sobel, : An isotropic image gradient operator. In H. Freeman, editor, Machine Vision for Three-Dimensional Scenes, pages 376–379. Academic Press, 1990.
8. J.M.S. Prewitt, : Object enhancement and extraction. In B.S. Lipkin and A. Rosenfeld, editors, Picture Processing and Psychopictorics. Academic Press, 1970.
9. D. Marr and E. Hildreth, : Theory of edge detection. IEEE Transactions on Pattern Analysis and Machine Intelligence, Proccedings of the Royal Society of London, vol. 207, pp. 187-217, 1980.
10. S. Mallat and S. Zhong, : Characterization of signals from multiscale edges. IEEE Trans. On PAMI. vol. 14, no. 7, (1992) 710–732
11. R. Deriche, : Using Canny's criteria to derive a recursively implemented optimal edge detector. Internat. J. Computer Vision. vol. 1, no. 2, (1987) 167-187
12. E. Bourennane and P. Gouton and M. Paindavoine and F. Truchetet, : Generalization of Canny-Deriche Filter for detection of noisy exponential edge. Signal Processing. vol. 82, (2002) 1317–1328
13. P. Gouton and H. Laggoune and Rk. Kouassi and M. Paindavoine, : Optimal Detector for Crest Lines. Optical engineering. vol. 39, no. 6, (2002) 1602–1611
14. D. Demigny, : On optimal linear filtering for edge detection, : IEEE Transactions on Image Processing. vol. 11, no. 7, (2002) 78–88

15. R. Missaoui, Sarifuddin M., N. Baaziz. and V. Guesdon, : Détection efficace de contours par d'ecomposition de l'image en régions homoges. 3rd International Conference: Sciences of Electronic, Technologies of Information and Telecommunications SETIT' 05. Sousse, Tunisia, March 2005.

3D Face Recognition using ICP and Geodesic Computation Coupled Approach

Boulbaba Ben Amor[1], Karima Ouji[2], Mohsen Ardabilian[1], Faouzi Ghorbel[2], and Liming Chen[1]

[1] LIRIS, Lyon Research Center for Images and Intelligent Information Systems, Ecole Centrale de Lyon. 36, av. Guy de Collongue, 69134 Ecully, France.
{boulbaba.ben-amor, mohsen.ardabilian, liming.chen} @ec-lyon.fr
[2] GRIFT, Groupe de Recherche en Images et Formes de Tunisie, Ecole Nationale des Sciences de l'Informatique, Tunisie.
{karima.ouji, faouzi.ghorbel} @ensi.rnu.tn

13.1 Introduction

Over the past few years, biometrics and particularly face recognition and authentication have been applied widely in several applications such as recognition inside video surveillance systems, and authentication within access control devices. However, as described in the *Face Recognition Vendor Test* report published in [1], as in other reports, most commercial face recognition technologies suffer from two kinds of problems. The first one concerns inter-class similarity such as twins'classes, and fathers and sons'classes. Here, people have similar appearances which make their discrimination difficult. The second, and the more important complexity, is due to intra-class variations caused by changes in lighting conditions, pose variations (i.e. three-dimensional head orientation), and facial expressions. On the one hand, lighting conditions change dramatically the 2D face appearance ; consequently approaches only based on intensity images are insufficient to employ. On the other hand, pose variations present also a considerable handicap for recognition by performing comparisons between frontal face images and changed viewpoint images. In addition, compensation of facial expressions is a difficult task in 2D-based approaches, because it significantly changes the appearance of the face in the texture image.

In this chapter we introduce a new face recognition/authentication method based on new face modality: *3D shape of face*. The remainder of the chapter is organized as follows: Section (2) reviews the recent progress in 3D face recognition research field. Section (3) describes an overview of the proposed approach. In Section (4), we focus on developed works for $2\frac{1}{2}$D vs. 3D face matching via *ICP*. The section (5) describes the geodesic computation technique for facial expression compensation. In section (6), we emphasize the evaluations of the developed method on $ECL - IV^2$ 3D face database.

13.2 Recent progress on 3D face recognition

Current state-of-the-art in face recognition is interesting since it contains works which aim at resolving problems regarding this challenge. The majority of these works use intensity faces'images for recognition or authentication, called 2D model-based techniques. A second family of recent works, known as 3D model-based, exploits three-dimensional face shape in order to mitigate some of these variations. Where some of them propose to apply subspace-based methods, others perform shape matching algorithm. Figure 13.1 present our vision and taxonomy for the face recognition techniques which can de categorized in four classes: 2D vs. 2D, 3D vs. 3D, multimodal 2D+3D, and 2D vs. 2D via 3D.

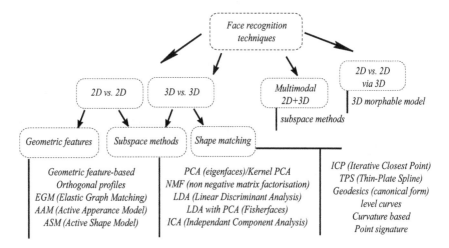

Fig. 13.1. A taxonomy of face recognition techniques.

As described in [6][9][8], classical linear and non-linear dimensional reduction techniques such as PCA and LDA are applied to range images from data collection in order to build a projection sub-space. Further, the comparison metric computes distances between the obtained projections. Shape matching-based approaches rather use classical 3D surface alignment algorithms that compute the residual error between the surface of probe and the 3D images from the gallery as already proposed in our works [13][14] and others as [11] and [3]. In [2], authors present a new proposal which considers the facial surface (frontal view) as an isometric surface (i.e. length preserving). Using a global transformation based on geodesics, the obtained forms are invariant to facial expressions. After the transformation, they perform one classical rigid surface matching and *PCA* for sub-space building and face matching. A good reviews and comparison studies of some of these techniques (both 2D and 3D) are given in [10] and [7]. Another interesting study which compares *ICP* and *PCA* 3D-based approaches is presented in [5]. Here, the authors show a baseline performance between these approaches and conclude that *ICP*-based method performs better than a *PCA*-based method. Their challenge is expression changes, particularly "eye lip open/closed" and "mouth open/closed".

In the present chapter, we discuss accuracy of a new 3D face recognition method using the *ICP*-based algorithm with a particular similarity metric based on geodesic maps computation. A new multi-view and registered 3D face database which includes full 3D faces and probe images with all these variations is collected in order to perform significant experiments.

13.3 Overview of the proposed 3D face matching method

Our identification/authentication approach is based on dimensional surfaces of faces. As illustrated by figure 13.2, we build the full 3D face database with neutral expressions. The models inside includes both shape and texture channels: *the off-line phase*. Second, a partial probe model is captured and compared to all full 3D faces in the gallery (if identification scenario) or compared to the genuine model (if authentication scenario): *the on-line phase*. The main goal of the availability of full 3D face models in the gallery is to allow comparison of the probe model for all view point acquisition.

The core of our recognition/authentication scheme consists of aligning then comparing the probe and gallery facial surfaces. The first step, approximates the rigid transformations between the presented probe and the full 3D face, a coarse alignment step and then a fine alignment step via *ICP* (Iterative Closest Point) algorithm [4] are applied. This algorithm is an iterative procedure minimizing the *MSE (Mean Square Error)* between points in partial model and the closest points in the 3D full model. One of the outputs of the algorithm result is two matched sets of points in the both surfaces. For the second step, two geodesic maps are computed for the pair of vertices in the matched surfaces. The recognition and authentication similarity score is based on the distance between these maps.

13.4 ICP for 3D surface alignement

One of interesting ways for performing verification/identification is the 3D shape matching process. Many solutions are developed to resolve this task especially for range image registration and 3D object recognition. The basic algorithm is the Iterative Closest Point developed by *Besl et al.*, in [4].

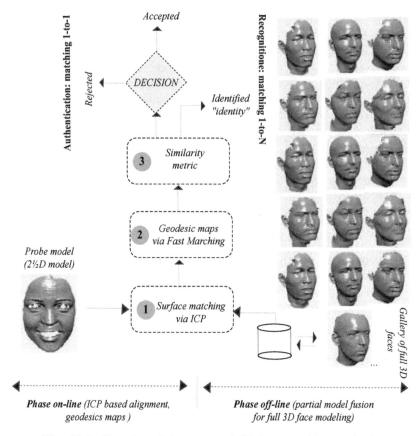

Fig. 13.2. Overview of the proposed 3D face matching method.

In our approach we consider first a coarse alignment step, which approximates the rigid transformation between models and bring them closer. Then we perform a fine alignment algorithm which computes the minimal distance and converges to a minima starting from the last initial solution. The last alignment step is based on this well-known Iterative Closet Point algorithm [4]. It is an iterative procedure minimizing the *Mean Square Error (MSE)* between points in one view and the closest vertices, respectively, in the other. At each iteration of the algorithm, the geometric transformation that best aligns the probe model and the 3D model from the gallery is computed. Intuitively, starting from the two sets of vertices $P = \{p_i\}$, as a reference data, and $X = \{y_i\}$, as a test data, the goal is to find the rigid transformation (R, t) which minimizes the distance between these two sets. The target of *ICP* consists in determining for each vertex p_i of the reference set P the nearest vertex in the second set X within the meaning of the Euclidean distance.

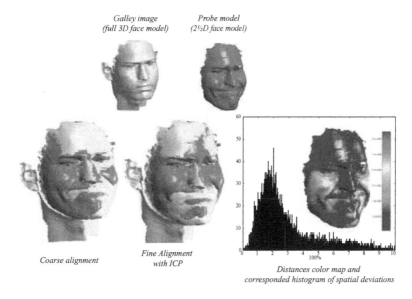

Fig. 13.3. ICP-based 3D surface alignement.

The rigid transformation, minimizing a least square criterion (1), is calculated and applied to each point of P:

$$e(R,t) = \frac{1}{N} \sum_{i=0}^{N} \|(Rp_i + t) - y_i\|^2 \qquad (13.1)$$

This procedure is alternated and iterated until convergence (i.e. stability of the minimal error). Indeed, total transformation (R,t) is updated in an incremental way as follows: for each iteration k of the algorithm: $R = R_k R$ and $t = t + t_k$. The criterion to be minimized in the iteration k becomes (2):

$$e(R_k, t_k) = \frac{1}{N} \sum_{i=0}^{N} \|(R_k(Rp_i + t) + t_k - y_i\|^2 \qquad (13.2)$$

The *ICP* algorithm presented above always converges monotonically to a local minimum [4]. However, we can hope for a convergence to a global minimum if initialization is good. For this reason, we perform the previous coarse alignment procedure before the fine one (cf. Figure 13.3).

13.5 Geodesic computation for 3D surface comparison

3D surface alignment via *ICP* does not have succeeded in curing the problem of facial expressions which present non-rigid transformations, not able to be modelled by rotations and translations. Thus, in a second stage, we propose to compute geodesic distances between pairs of points on both probe and gallery facial surfaces since this type of distances is invariant to both rigid and non-rigid transformations, as concluded in [2]. Therefore, an efficient numerical approach called the *fast marching method* [16] is applied for geodesic computations. A geodesic is a generalization of straight line notion into curve spaces [15]. A geodesic is the shortest path between two points on the considered surface (as shown by figure 13.4).

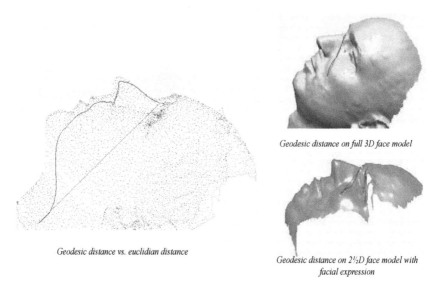

Geodesic distance on full 3D face model

Geodesic distance vs. euclidian distance

Geodesic distance on 2½D face model with facial expression

Fig. 13.4. Geodesic distance vs. euclidian distance computations on full 3D and partial face models with facial expressions.

13.5.1 Fast Marching on triangulated domains

The *fast marching method*, introduced by *Sethian* [15] is a numerically consistent distance computation approach that works on rectangular grids. It was extended to triangulated domains by *Kimmel & Sethian* in [16]. The basic idea is an efficient numerical approach that solves the *Eikonal* equation $|\nabla u| = 1$, where at the source point s the distance is zero $u(s) = 0$, namely. The solution u is a distance function and its numerical approximation is computed by a monotone update scheme that is proven to converge to the *'viscosity'* smooth solution.

The idea is to iteratively construct the distance function by patching together small plans supported by neighboring grid points with gradient magnitude that equals one. The distance function is constructed by starting from the sources point, S, and propagating outwards. Applying the method to triangulated domains requires a careful analysis of the update of one vertex in a triangle, while the u values at the other vertices are given. For further details in this theory, we refer to [16].

13.5.2 Application to 3D face recognition

After $2\frac{1}{2}$D vs. 3D *ICP* alignment step, we propose to compute geodesic maps on overlapped surfaces in both probe and gallery facial surfaces. We dispose of a list of the corresponding vertices already provided by *ICP*. We consider for more significantly and robust computation the zones limited by a pair of vertices in probe model and 3D full face model. As shown in figure 13.5 (C) the source vertex in probe model is noted S_1, the limit vertex L_1 and their correspondants in 3D face model: the source vertex S_2 and the limit vertex L_2 (see figure 13.5 (B)). Computing geodesic distances in the zone limited by these vertices is less sensitive to errors. In fact, the mouth region can present a hole if the mouth is opened, this introduces errors in geodesic computation algorithms, as shown by figure 13.5 (A).

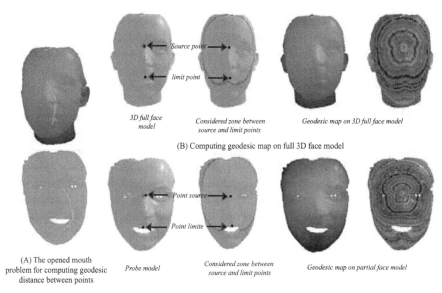

(A) The opened mouth problem for computing geodesic distance between points

(B) Computing geodesic map on full 3D face model

(C) Computing geodesic map on probe model (2½D model)

Fig. 13.5. Geodesic maps computation: (A) the opened mouth problem, (B) computing geodesic map on fill 3D face model, and (C) computing geodesic map on partial face model.

We propose to calculate a geodesic distance map on the $2\frac{1}{2}$D probe model via the extension of *fast marching method* to triangulated domains proposed by *Sethian & Kimmel* [16]. In fact, the algorithm starts on the source vertex S_1 and propagates along all the facial surface saving on every met vertex the geodesic distance which separate him to the source vertex S_1. All these distances make up the vector V_1 (the first geodesic map). Each line of V_1 contains the geodesic distance separating S_1 and the vertex having the index i of the $2\frac{1}{2}$D mesh. Then, we compute geodesic distance map on the 3D mesh with the same principle. In this case, the source vertex is a vertex S_2 and geodesic distance map is a vector V_2 . Each line of V_2 contains the geodesic distance separating S_2 from the vertex of the 3D mesh corresponding to the vertex having the index i of the 3D mesh. Finally, we compute the vector V as $V[i] = |V_2[i] - V_1[i]|$ and the similarity metric is the standard deviation of V and which we use for the recognition process.

In our implementation, we consider only vertices which are situated above mouth. In other words, V_1 and V_2 contain only geodesic distance from source point to points in the higher part of the face. In fact, we compute all geodesic distances on facial surface $2\frac{1}{2}$D and we get rid of all distances which value is more than the distance separating S_1 from L_1. Moreover, we compute all geodesic distances on 3D facial surface and we get rid of all distances which value is more than the distance separating S_2 and L_2. We justify our choice by the fact that if the probe person is laughing for example, the $2\frac{1}{2}$D mesh contains a hole in mouth. Thus, all geodesic distances computed for vertices situated under the lower lip are very different from geodesic distances of their correspondants on the 3D mesh which have no hole in mouth as illustrated by figure 13.5.

13.6 Experiments and future works

In this section we present some experimental results of the presented approach performed on $ECL - IV^2$ 3D face database. Figure 13.6 illustrates this rich database which contains about 50 subjects (about 50 full 3D face and 400 probe models [13]). It includes full 3D face gallery (figure 13.6 (A)), and eight partial models including variations in facial expressions, poses and illumination (figure 13.6 (D)).

We produce for each experiment, labelled by the considered variation, both the error trade-off curve (for authentication scenario) and the rank-one recognition rates (for identification scenario). Figure 13.7 presents the recognition rates for the elementary experiments and the global one (all probe images are considered). It is shown that less the facial deformation is, more invariant the method is. In addition, the proposed paradigm is invariant to both illumination and pose variation problems (see figure 13.8). This is done by, respectively, cancelling texture data in our processing scheme and the availability of the full 3D models in the gallery dataset. The *global Recognition Rate (RR)* is equal to 92.68% and the *Equal Error Rate (EER)* about 12% .

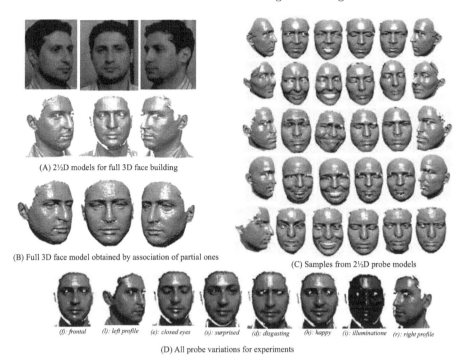

(A) 2½D models for full 3D face building

(B) Full 3D face model obtained by association of partial ones

(C) Samples from 2½D probe models

(f): frontal *(l): left profile* *(e): closed eyes* *(s): surprised* *(d): disgasting* *(h): happy* *(i): illuminatione* *(r): right profile*

(D) All probe variations for experiments

Fig. 13.6. $ECL - IV^2$: new multi-view registred 3D face database.

Experiments	(d)	(e)	(f)	(h)	(i)	(l)	(r)	(s)	(all)
rank-one rate (%)	80.48	92.68	97.56	90.24	97.56	97.56	97.56	87.80	92.68

Legend: *(d)* disgusting, *(e)* closed eyes, *(f)* frontal with neutral expressions, *(h)* happy, *(i)* uncontrolled illumination, *(l)* left profile, *(r)* right profile, *(s)* surprise, and *(all)* all probe images.

Fig. 13.7. Rank-one recognition rates for elementary experiments and global one.

This approach shows more significant rates in absence of expression, this is because in presence of expression, the ICP-based alignment method provides less accurate matched points. In order to surmount this problem, it is more significant to perform alignment only on static regions of the face. In our future work, we plan to combine the proposed approach with our region-based method for mimics segmentation already presented in [12]. In this approach the alignment only concerns the static regions of the face.

In conclusion, the presented approach is invariant to pose and illumination variations. However, it shows some limits in presence of important facial expressions. As described in the last paragraph, it can be enhanced by performing alignment and geodesic computation on only static region of the face (cf. Figure 13.9).

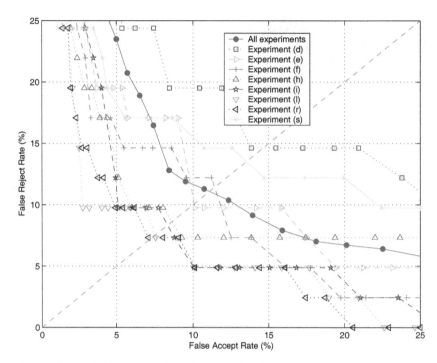

Fig. 13.8. DET (Error trade-off) curves for elementary and all experiments.

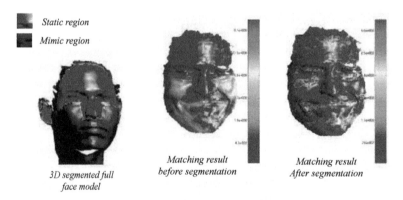

Fig. 13.9. Region-based ICP for more significant 3D face alignement and matching.

Acknowledgments.

The authors thank the Biosecure participants for their contribution to collect this 3D face database and their significant remarks. Authors thank also Gabriel Peyre, who provides us his implementation of the Fast Marching algorithm.

References

1. P.J. Phillips, P. Grother, R.J Micheals, D.M. Blackburn, E. Tabassi, J.M. Bone: Technical report NIST. FRVT 2002: Evaluation Report (Mars 2003).
2. A. M. Bronstein, M. M. Bronstein, R. Kimmel: Three-dimensional face recognition. In: International Journal of Computer Vision (IJCV): pages 5-30 (august 2005).
3. Xiaoguang L. and Anil K. Jain: Integrating Range and Texture Information for 3D Face Recognition. In: Proc. 7th IEEE Workshop on Applications of Computer Vision pages 156-163 (2005).
4. Paul J. Besl and Neil D. McKay: A Method for Registration of 3-D Shapes. In: Proc. IEEE Trans. Pattern Anal. Mach. Intell vol. 14 pages 239-256 (1992).
5. Chang, K. J. and Bowyer, K. W. and Flynn, P. J.: Effects on facial expression in 3D face recognition. Proceedings of the SPIE, Volume 5779, pp. 132-143 (2005).
6. Gang Pan and Zhaohui Wu and Yunhe Pan: Automatic 3D Face Verification From Range Data. Proc. IEEE International Conference on Acoustics, Speech, and Signal Processing, pp. 193-196 (2003).
7. Chenghua Xu and Yunhong Wang and Tieniu Tan and Long Quan: Depth vs. Intensity: Which is More Important for Face Recognition?. Proc. 17th International Conference on Pattern Recognition, 2004.
8. Sotiris Malassiotis and Michael G. Strintzis: Pose and Illumination Compensation for 3D Face Recognition. Proc. IEEE International Conference on Image Processing, pp 91-94, 2004.
9. Thomas Heseltine and Nick Pears and Jim Austin: Three-Dimensional Face Recognition: An Eigensurface Approach. Proc. IEEE International Conference on Image Processing, pp 1421-1424, 2004.
10. Kyong I. Chang and Kevin W. Bowyer and Patrick J. Flynn: An Evaluation of Multi-modal 2D+3D Face Biometrics. IEEE Transactions on PAMI, vol. 27 pp 619-624, 2005.
11. Charles Beumier and Marc Acheroy: Automatic 3D Face Authentication. Image and Vision Computing, vol. 18 pp 315-321, 2000.
12. B. Ben Amor and M. Ardabilian and L. Chen: Enhancing 3D Face Recognition By Mimics Segmentation. Intelligent Systems Design and Applications (ISDA'2006), pp. 150-155, 2006.
13. B. Ben Amor and M. Ardabilian and L. Chen: New Experiments on ICP-based 3D face recognition and authentication. Proceeding of International Conference on Pattern Recognition (ICPR'2006), pp. 1195-1199, 2006.
14. B. Ben Amor and K. Ouji and M. Ardabilian and L. Chen: 3D Face recognition by ICP-based shape matching. Proceeding of IEEE International Conference on Machine Intelligence (ICMI'2005).
15. J.A. Sethian: A Fast Marching Level Set Method for Monotonically Advancing Fronts. Proc. Nat. Acad. Sci, 93, 4, 1996.
16. J.A. Sethian et R. Kimmel: Computing Geodesic Paths on Manifolds. Proc. Natl. Acad. Sci., 95(15):8431-8435, 1998.

Part III

Face Recognition and Shape Analysis

A3FD: Accurate 3D Face Detection

Marco Anisetti[1], Valerio Bellandi[1], Ernesto Damiani[1], Luigi Arnone[2], and Benoit Rat[3]

[1] Department of Information Technology, University of Milan
via Bramante, 65 - 26013, Crema (CR), Italy
{anisetti, bellandi, damiani}@dti.unimi.it
[2] STMicroelectronics Advanced System Research group
Agrate Brianza (MI), Italy
luigi.arnone@st.com
[3] EPFL Ecole Polytechnique Federale de Lausanne
Lausanne, Swiss
benoit.rat@epfl.ch

Summary. Face detection has recently received meaningful attention, especially during the past decade as one of the most prosperous applications of image analysis and understanding. Video surveillance is for example, one emerging application environment. This chapter presents a method for accurate face localization through a coarse preliminary detection and a following 3D refinement (A3FD: Accurate 3D Face Detection). A3FD can be useful applied to video surveillance environment thanks to the performance and the quality of the results. In fact for many application (e.i. face identification) the precision of face features localization is a real critical issue. Our work is therefore focused on improving the accuracy of the location using a 3D morphable face model. This technique reduces the false positive classification of a face detector and increases the precision of the positioning of a general face mask. Our face detection system is robust against expression, illumination and posture changes. For comparison purposes we also present some preliminary results on largely used face database.

14.1 Introduction

Images containing faces are essential to intelligent vision-based human computer interaction and applications. Nevertheless current recognition systems have reached a certain level of maturity (for a complete overview see [1]), their success is limited by the conditions imposed by many real applications. For instance, recognition of face images acquired by a surveillance system in an outdoor environment with changes in illumination and/or pose and expression, still remains a challenge task. In other words, the applicability of current face detection system in real application is confined in some particular cases. Furthermore, many detection systems do not reach the level of accuracy required by many real applications especially in terms of precision when it comes to face location. For instance, it is known that face identification systems produce a more accurate result when working on normalized faces. Our previous works demonstrate also that normalization across illumination, posture and expression could improve the quality even with a simple eigenface classifier [2]. Summarizing the accuracy of the detection system has a comparable or even a greater importance for real application. Furthermore, considering that real application relies on camera presence, we need to manage video streaming instead of static images. Considering this environment, the percentage of detection in a single image has little importance in comparison with the location quality. In fact we rely on the fact that, in a reasonable period of time the face presents a scene which could be in a situation that permits to be detected (in fact the principal causes of missed faces is related to situational causes: posture illumination expression for example). Considering these further hypotheses, we focused our attention on issues coupled with the accuracy of face location (in a precise localization sense and in terms of posture and face morphology) instead of detection quality (face detection vs face presentation in the images).

Our contributions in face detection problem are: i) Accurate 3D posture evaluation based on our steepest descent approach, ii) Independence from illumination and expression changing. We do not focus our attention on improving the percentage of correct detection; we use an efficient mixture of well known techniques as first layer detector, (this brings a high true positive detection rate but also high false positive ones) and an AdaBoost based classification, that strongly reduce the false positive. We showed that we can refine the face location and further reduce the false positive detection, using a technique taken from our past experiences in face tracking algorithm. In this context, the two largely used techniques are based on Active Appearance Model (AAM) [3] both 2D and 3D and morphable models (MM)[4]. In these works concerning face tracking, there are no references to detecting techniques; meanwhile, in this chapter we also try to investigate in order to obtain faster techniques to extract the relevant information indispensable for refinement algorithm. Furthermore, our refinement algorithm relies on different approaches, instead of the AMM or MM making it more suitable for real application (for conciseness, we refer to a detailed description of the differences with our tracking approaches in our previous works [5] [6]). In these previous works, we need several points manually selected on a face, in order to detect the initial location (posture, morphology and expression) as a starting point for our tracking algorithm. In this chapter, we present an approach based on 5 points that can be quickly detected by our automatic system (our preliminary face detector) and that do not required an high precision thanks to the use of our 3D refinement system that estimates the other morphological and posture and expression parameters. This approach automates the accurate face detection obtaining comparable results in terms of quality respect to the manually selected points approach.

The remaining part of the chapter is organized as follows. Section 14.2 describes the Face Detection and Features Selection algorithms that localize face and all relevant features for preliminary posture evaluation. Section 14.3 presents an accurate posture and morphology detection approach including robustness applied to expression and illumination parameters. Section 14.4 includes a conclusion and a discussion completing the chapter.

14.2 Preliminary face detection

In this section we underline the main characteristic of our preliminary face detection strategy. As pre-described in Section 14.1, our face detection is mainly composed by the following components: i) preliminary face detection, ii) accurate localization and posture detection. Therefore, in this section we focus our attention on the first component. This component's goal is to detect probable faces and localize few face features useful for posture evaluation. At this stage, we do not need to take into account the percentage of correct detection or precision of the feature location, because we delegate the refinement to the accurate localization component (Section 14.3). We also need to have a fast detection because we focus our attention on video sequences instead of still images and on real time application in the surveillance environment. Our face detection algorithm works with not lateral face, and with partial occlusion. We test our approach with several video databases (Hammal-Caplier database [7], MMI Facial Expression Database collected by M. Pantic & M.F. Valstar [8]) and AR database [9] that contain images with different type of illumination and occlusion. Our preliminary detection algorithm relies on the following steps:

- *Adaptive Skin Detection*: Aimed to detect the blobs that include faces. Our adaptation strategy of skin model permits to cover most changes in illumination (when in white light source).
- *Local feature selection and validation*: To detect some meaningful facial features inside the skin blobs (i.e. eyes positions). Next we perform a validation using AdaBoost classification with Haar features like in [10].
- *Face Classification*: We perform localization, normalization and classification using Adaboost approach.

The main idea of this multi-step algorithm is to obtain an coarse-to-fine classification. Taking into consideration this idea the classification starts from rough to precise face detection, with the correspondent growing levels of reliability of correct classification. Following we describe each part in detail.

14.2.1 Adaptive Skin detection

This steps aim is to reduce the search area the skin-map approach. This approach doesn't focus on precise detection of face contours. It only attempts to detect areas that include possible faces. Our skin detection strategy works with one gaussian [4] on $YCbCr$ color space. For every pixel x of the images we obtain the skin probability (or rather what belongs to the skin's region defined by the gaussian). Using a threshold on this probability we obtain the skin blobs. This type of skin-map often suffers illumination problems and changing in skin colour due racial belongings. To reduce this type of disturbance we train our gaussian using different racial skins. Respecting the illumination issue, only taking in consideration white light and the digital camera's white compensation, we perform what we call an adaptive skin operation. To perform this adaptation, we arrange the skin mean value adjustment in this manner: i) compute the skin-map, ii) dilate the skin-map region, iii) recompute the mean of $CbCr$ gaussian considering also the color of the pixel under the new enlarged region. This process could produce an enlargement of the skin region that can include some non face areas. This is not a problem since our goal is not to detect the perfect face region but only to reduce the areas of face searching in respect to an entire frame. The main advantage is that in many cases this adaptation compensates a light change (Fig. 14.1). From our experience, even though this skin colour adjustment enlarges the blobs area, the eye region (even for a few pixel) still remains out of the skin-map region. We then perform some simple morphological operations to obtain a more precise region of skin and non skin region inside a skin blob (Fig. 14.1).

To conclude with, we perform our first area detection level that contains a probable face. Some other strategies could be used like ellipse searching on a gradient map and so forth. Our implementation permits to save computational power reaching a speed of 5 frames per second in a Matlab ®implementation (640x480 frame size on 2 Ghz Pentium IV).

14.2.2 Local feature selection and validation

After the rough preliminary facial area localizations, we obtain the image regions where a possible face can be located. These very rough localizations do not depend upon the face postures. At this stage we do not therefore restrict our face detection strategies to only frontal face [5] according to our coarse-to-fine and progressively selective strategy. To find the face inside the skin region we use the non-skin areas included in every skin blobs. Each non-skin area is our possible eye location. The main idea is to detect a pair of non-skin regions that can possibly be the eyes of a face. This searching is performed intelligently to reduce the possible pair candidates with some heuristics based on the relative blobs' positions. The eyes searching process is performed as follows: i) Computing every probable pairs of non-skin region, ii) Computing

---

[4] We do not use a mixture of gaussian because we do not need to reach a great precision but we need to save computational time

[5] Using other approach like [10] a complete scanning of the images must be carried out and only face with posture in accord with training set can be detected.

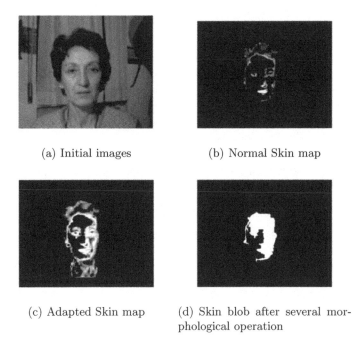

(a) Initial images   (b) Normal Skin map

(c) Adapted Skin map   (d) Skin blob after several mor-
                               phological operation

Fig. 14.1. Skin map process for blob detection.

the centroid of every non-skin region for each pairs, iii) Computing the affine transformation for normalizing the box containing the eyes. This normalized region is useful for the classification process. Fig. 14.2 shows some probable eye regions after the affine normalization and the relative face detected using AR database.

Fig. 14.2. Some example of probable pairs of eyes after normalization and the relative face after the eyes classification with AdaBoost in AR database.

We build a large training set with positive and negative examples. We used the eyes pairs images extracted with our algorithm and we labelled them manually. Generally speaking any kind of machine-learning-like approaches could be used to learn a classification function (for instance neural network or support vector machines). We chose to follow the approach present in the paper of Viola et al.[10] where some computed efficient Haar features associated with each image are used. Even though each feature can be computed very efficiently, computing the complete set is prohibitively expensive. The main challenge is to find these features. A very interesting hypothesis is formulated by Viola and Jones: a very small number of these features can be combined to create an effective cascade classifier using a variant of Adaboost. This cascade classifier is composed in different stages.

Each stage is called a strong classifier and it is trained by the AdaBoost algorithm with the following rules: To reject 60% of negative items and to insure that we detect at least 99.8% of eye pairs. Our first stage uses only 3 features. For the following stages, the complexity of detection and the number of features needed increase. Using the cascade permits to quickly drop several non-eye pairs in the first stages and very few eye pairs. At this step we do not need to reach a great reliability in the classification so we decide to use only a two stage AdaBoost for speeding-up every operation. In our experiment this classifier permit to discard the 89% of false eye pairs with no significant percentage of correct eye pairs discarded. Fig. 14.3 shows some detected faces in a image data base build with famous people found on internet.

Fig. 14.3. An example of a face detected with posture or even expression different from the frontal neutral view. Note that this face is already normalized in terms of tilting.

After the eye detection step, several simple operations are performed obtaining some crucial points for the eyes and mouth (5 points: the inner and outer corner of the eyes and the center of the upper lip). These points are quite stable even with expression changes. However the precision of these points is not critical at this level of the process thanks to our refinement estimation that we perform at the last stage.

14.2.3 Face normalization and classification

For every pair of eyes we reconstruct a face using the human face morphology ratios based on strategies and using the detected mouth points where available. Than we use a similar strategy of normalization adopted by eyes classification but using a more reliable AdaBoost classifier with several stages. The goal of this classification is to increase the reliability of the face detection system. This degree of reliability is estimated observing at which stage a candidate's face is rejected. More the candidate passes stages, greater is the reliability degree. In the table 14.1 we present the detection results on the AR data base. As it can noticed the correct detection results on images with black sunglasses is not present at all, and for images exposed to a yellow illuminant are really poor. This because we worked on the hypothesys of white illuminant and visible eyes.

Summing up, in order to keep the false positive rate of 0.28% we paid with a rate of true positive 72.9% without considering black sunglasses and yellow illuminat images. The latest refinement is performed with our 3D face model fitting system. This will increase much the rate of the correct recognition. When this last process does not converge to the end, the face will be classified as a probable face but not with an accurate estimated morphology position.

PhotoIndex	N. CorrectDetection	N. Images	Characteristic
1	99	135	
2	113	135	Smile
3	102	135	
4	104	135	OpenMouth, CloseEyes
5	24	135	YellowLightRight
6	14	135	YellowLightLeft
7	4	135	YellowLight
11	70	135	Scarf
12	11	135	Scarf, YellowLightRight
13	4	135	Scarf, YellowLightLeft
14	92	119	
15	102	120	Smile
16	93	120	
17	93	120	OpenMouth, CloseEyes
18	21	120	YellowLightRight
19	14	120	YellowLightLeft
20	1	120	YellowLight
24	61	120	Scarf
25	11	120	Scarf, YellowLightRight
26	9	120	Scarf, YellowLightLeft

Table 14.1. AR database correct recognition results.

14.3 Accurate location refinement via 3D tracking approach

To refine the quality of face location we use an extended version of our tracking algorithm [6]. To cover the problem of precise face location, we adopt the same 3D model and convergence strategy than our tracking algorithm considering the information extracted from face detection as a rough posture and morphological parameters initialization. This type of initialization is different from the our previous work when we compute the precise face parameters (posture, morphological and expression) using several points selected by hand, since it is difficult to detect these points automatically with the great precision required. So in real application this strategy can not be applicable. This is the main problem that we solved using a rough posture and morphological initialization and when refining the parameter using convergence algorithm. To perform this refinement we include the morphological basis used in our previous point based initialization, in the minimization coupled with the posture and expression parameters.

In terms of appearance, the texture of the subject is not extracted using the initialization parameters, since that the rough initialization do not provide an accurate texture. So we decided to extend our tracking algorithm including texture appearance parameters such as in AAM or MM. We compute these parameters using eigenface techniques on a face database properly normalized using our 3D mask (see [2] for detailed description) and include the first four principal eigenface in the linear appearance part (see formula 14.2) of our formulation in the same manner as the illumination parameters. This means that we also reconstruct the texture of the face using a linear combination of eigenface, and that this can be performed in the same way as the illumination. Summarizing we include morphological and appearance parameters in our minimization techniques obtaining a technique to refine a rough localization. This technique works well under the constraint that the initial rough initialization needs to be not far from the real posture, at least for translation and tilting. Furthermore, to perform the convergence, the frame image needs to be blurred for eliminating the initial difference between real face and the reconstructed ones (we perform a gaussian pyramid of blurring). For conciseness, we refer to our previous work [6] for complete mathematical dissertation over our minimization. Following we present the aforesaid adaptation of our tracking algorithm.

Summarizing, our minimization process relies on the presence of a face model. Like in our tracking approach, this face model includes a basis of movement to cover deformation due to expression and morphological aspect. We also include an appearance set of parameters obtained using eigenface approach. Than we have an average face and the four principal components. As in the eigenface approach, a linear combination of this eigenface can produce a certain appearance. Considering this idea, we can introduce in our tracking algorithm the eigenfaces as the appearance parameters using a linear appearance strategy.

Our goal is therefore to obtain posture, morphological estimation parameters and Action Units [12] deformation parameters in one minimization process between one frame and a average template of face T_{avg} plus the linear combination of eigenface (that we call $T_{avg+eig}$) that produce the most accurate approximation of face. We rely on the idea that face 2D template $T_{avg+eig}(x)$, will appear in the actual frame $I(x)$ albeit warped by $W(x;p)$, where $p = (p_1, \ldots, p_n, \alpha_1, \ldots, \alpha_m)$ is vector of parameters for 3D face model with m parameters (morphological and Action Units for expression) and x are pixel coordinate from image plain. Thanks to this assumption, we can obtain the movement and morphological-expression parameter p by minimizing function (14.1). If $T_{avg+eig}(x)$ is the template with correct pose morphology and expression p and $I(x)$ is the actual frame, assuming that the image do not differs to much respect with reconstruct $T_{avg+eig}$ and that the preliminary detection posture has the satisfactory precision, the correct pose and morphology and expression p can be obtained by minimizing the sum of the square errors between $T_{avg+eig}$ and $I(W(x;p))$:

$$\left(\sum_x [I(W(x;p)) - T_{avg+eig}]^2 \right) \tag{14.1}$$

For this minimization we use the Lucas-Kanade approach [13] with forward additive implementation [14].

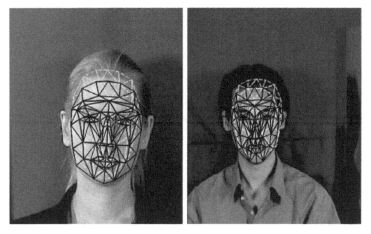

Fig. 14.4. Comparisons between initialization with preliminary estimation (white mask), and after the minimization (black line) on MMI database.

Now to include the estimation of eigenface parameters inside our minimization algorithm, we consider the problem as a linear appearance model so the $T_{avg+eig}$ becomes only the average face T_{avg} and the eigenface parameters are put inside the appearance factor. As occurs for illumination basis:

$$\sum_x [I(W(x;p)) - T(x) - \sum_{i=1}^{m} \lambda_i A_i(x) - \sum_{j=1}^{n} \sigma_j Eig_j(x)]^2 \tag{14.2}$$

where A_i with $i = 1, \ldots m$ and Eig_j with $j = 1, \ldots n$ is a set of known appearance variation images and λ_i and σ_j with $i = 1, \ldots m$ and $i = 1, \ldots n$ are the appearance parameters. Global illumination change can be modelled as described in our previous work and the changing in texture due the appearance of the face can be also modelled as linear combination of computed eigenface with average face. So thanks to linear appearance variations techniques, this function can be also minimized using a Lucas-Kanade like approach. Concluding, the convergence of this process brings further information upon the reliability of face detection. In fact in case of false positive our approach do not converge or if it converge, it converges to unreliable parameters. From our preliminary test, this occurs for a 97% using AR and MMI database and with 89% over more complex our own database. Concluding this last refining step of the detection algorithm, demonstrates some interesting promising results in terms of precision. In fact we perform a further comparison using our previously developed system, that use several manually placed points, and we obtain that the difference in terms of posture, expression and morphology parameters estimated is still confined. This results shows that the proposed strategy obtains a comparable performance in terms of precision. This results shows that the proposed strategy obtains incomparable performance in terms of precision. This is a promising result for the correct functioning of the face tracking system. Furthermore thanks to the last detection step, we have an increase in the correct classification due to the reducing of the false positive classification rate. These preliminary results will be verified in the close future on a more extensive test using various database.

14.4 Conclusion and discussion

The main characteristics of this 3D morphable face model-based detection algorithm can be sum up as:

- Reducing the percentage of false positive classification.
- Increasing the precision in the positioning of the face mask. This is not fundamental just for tracking algorithms, but also for the classification algorithms based on appearance and consequently on the correct normalization of the face.
- The minimization process in the last step of the detection, extract not only the position of the face, but also some morphological and appearance parameters (eigenfaces coefficients) that can be used directly as feature for an identifier. We showed in a past work [2] how the normalized face based eigenfaces coefficients can ease the classification process. In this case we can combine some morphological features (related to the face morphology) to the eigenfaces parameters. This can increase the recognition rate of an identification block.

Next step will be an exhaustive testing of the whole system on the same real cases, but on bigger surveillance oriented databases. We have planned also to develop the identification system based on the parameters extracted from this detection process.

References

1. M. Yang, D. Kriegman, and N. Ahuja, "Detecting face in images: A survey," *IEEE Transactions on Pattern Analysis and Machine Intelligence*, vol. 24, no. 1, pp. 34–58, 2002.
2. F. Beverina, G. Palmas, M. Anisetti, and V. Bellandi, "Tracking based face identification: A way to manage occlusion, and illumination, posture and expression changes," in *Proc. of IEE 2nd International Conference on Intelligent Environments IE06*, Athens, Greece, 2006.
3. T. Cootes, G. Edwards, and C. Taylor, "Active appearance mode," *IEEE Transactions on Pattern Analysis and Machine Intelligence*, vol. 23, no. 6, pp. 681 – 685, Jun. 2000.
4. V. Blanz and T. Vetter, "Face recognition based on fitting a 3d morphable model," *IEEE Transactions on Pattern Analysis and Machine Intelligence*, vol. 25, no. 9, pp. 1063 – 1074, 2003.
5. E. Damiani, M. Anisetti, V. Bellandi, and F. Beverina, "Facial identification problem: A tracking based approach," in *IEEE International Symposium on Signal-Image Technology and InternetBased Systems (IEEE SITIS05)*.
6. V. Bellandi, M. Anisetti, F. Beverina, and L. Arnone, "Face tracking algorithm robust to pose, illumination and face expression changes: A 3d parametric model approach," in *Proc. of International Conference on Computer Vision Theory and Applications (VISAPP)*, Setubal, Portugal, Feb. 2006.
7. Z. Hammal, A. Caplier, and M. Rombaut, "A fusion process based on belief theory for classification of facial basic emotions," *Proc. Fusion'2005 the 8th International Conference on Information fusion (ISIF 2005)*, 2005.
8. M. Pantic, M. Valstar, R. Rademaker, and L. Maat, "Web-based database for facial expression analysis," *Proc. IEEE Int'l Conf. Multmedia and Expo (ICME'05), Amsterdam, The Netherlands*, 2005.
9. A. Martinez and R. Benavente, "The ar face database," CVC Technical Report, Tech. Rep. 24, June 1998.
10. P. Viola and M. Jones, "Robust real-time object detection," *Second Intl.Workshop on Stat. and Comp. Theories of Vision*, 2001.
11. "http://www.navyrain.net/facedetection.html," Tech. Rep.
12. P. Ekman and W. Friesen., "Facial action coding system: A technique for the measurement of facial movement." *Consulting Psychologists Press*, 1978.
13. B. Lucas and T. Kanade, "An iterative image registration technique with an application to stereo vision," *Proc. Int. Joint Conf. Artificial Intelligence*, pp. 674–679, 1981.
14. S. Baker and I. Matthews, "Lucas-kanade 20 years on: A unifying framework," *International Journal of Computer Vision*, vol. 56, no. 3, pp. 221 – 255, February 2004.

15

Two dimensional discrete statistical shape models construction

Isameddine Boukhriss, Serge Miguet, and Laure Tougne

Lyon2 University, LIRIS Laboratory
Batiment C, 5 av. Pierre Mendes-France
69676 Bron Cedex, France
{isameddine.boukhriss,serge.miguet,laure.tougne}@univ-lyon2.fr

15.1 Introduction

Models of shape are used widely in computer vision. They are of great interest in objects localization, tracking or classification. Many objects are non-rigid, requiring a deformable model in order to capture their shape variability. One such model is the Point Distribution Model (PDM). An object is modeled in terms of landmark points positioned on contours. By identifying such points on a set of training examples, a statistical approach (principal component analysis, or PCA) can be used to compute the mean object shape, and the major modes of shape variations. The standard PDM is based purely on linear statistics (the PCA assumes a Gaussian distribution of the training examples in shape space). For any particular mode of variation, the positions of landmark points can vary only along straight lines. Non-linear variation is achieved by a combination of two or more modes. Our approach encodes shapes with a set of particular points resulting from a polygonalization process. This guarantees the reversibility, in the construction process of the final model. However, shapes could have different numbers of points encoding their contours, which is not convenient for the PCA. This problem is solved with a controlled matching process which ensures that for every point of a given shape contour, we could find a destination point in any other contour. This procedure is explained in 15.3. In section 15.2, we show how we encode shapes contours and how we compute their discrete parameters. Section 15.4 explains how the final model is extracted. But first of all, let us talk about previous works about shape models.

In [20], we focused on shape matching and recognition, but in this chapter we extend the approach to extract deformable models from a set of shapes. Many approaches are proposed for shape models construction. For studying shape variations, these approaches could be 'hand-craft', semi-automatic or automatic. Yuille and al [5] build up a model of a human eye using combinations of parameterized circles and arcs. This can be effective, but it is complicated, and a completely new solution is required for every new case. Staib and Duncan [6] represent shapes using fourier descriptors of closed curves. The complexity depends on coefficients and the generalization is not possible to open contours. Kass et al [7] introduced Active Contour Models (snakes) which are energy minimising curves. When no model is imposed, they are not optimal for locating objects which have a known shape. Alternative statistical approaches are described by Goodall [8] and Bookstein [9], they use statistical techniques for morphometric analysis.

The most common modelling variability approach is to allow a prototype to vary according to some learned model. Park and al. [10] and Pentland and Sclaroff [11] represent the outlines or surfaces of prototype objects using finite element methods and describe variability in terms of vibrational modes. Turk and Pentland [12] use principal component analysis to describe the intensity patterns in face images in terms of a set of basis functions, or 'eigenfaces'. Poggio and Jones [13] synthesize new views of an object from a set of example views. They fit the model to an unseen view by a stochastic optimization procedure. In [14], computing a curve atlas, based on deriving a correspondence between two curves, is presented. The optimal correspondence is found by a dynamic programming method based on a measure of similarity between the intrinsic properties of curves. Our approach has a common part with their technique in the matching process shown in section 15.3 which is a pre-stage to statistical analysis and model computing. But our technique is discrete, it does not subdivide uniformly contours and optimize correspondence between two sets of points with different sizes. The extraction and characterization of these points is explained in the following section. The section 15.3 deals with the process of matching between shapes. When this process is optimized on all shapes, we extract their statistical representative in section 15.4.

15.2 Data adjustment

Let us consider two binary figures F and F', each of them contains only one connected component, without hole, called shape in the following and denoted respectively by f and f'. We proceed by a registration and scaling process. f and f' are aligned according to their maximal elongation (principal component associated to the first eigen vector) with maximum interior intersection. These operations are based on moments of order two computed for each shape, they produce ellipses of better approximation describing shapes orientation and describe dispersion of data.

Let us now consider the respective borders of each shape, that is to say the set of points that belongs to the shape and that have at least one 4−background neighbor. Note that we consider then 8−connected borders. Let us denote respectively by C and C' the borders of f and f'. We suppose that the border C contains n points and the border C' contains M points with N not necessary equals to m. These points are the extremities of the segments obtained by Debled's linear polygonalization algorithm [1]. Let us denote by P_i with i from 0 to n the extremities of segments extracted from C, remark that $P_0 = P_n$, and P'_j with j from 0 to m the extremities of segments extracted from C' with $P'_0 = P'_m$. On each point P_i and P'_j we compute parameters that will help to associate a cost to each pair of points (P_i, P'_j).

The first parameter to compute is the distance between the points of one pair. So, we consider the Euclidean distance:

$$d(P_i, P'_j) = \sqrt{(x_{P_i} - x_{P'_j})^2 + (y_{P_i} - y_{P'_j})^2}$$

The two other parameters we compute compare locally the two borders C and C'. One measures the difference between the curvatures in the points P_i and P'_j and the second compares the normal vectors.

To the pair of points (P_i, P'_j) we associate the difference of curvatures :

$$\kappa(P_i, P'_j) = \mid \kappa_{P_i} - \kappa_{P'_j} \mid$$

In order to avoid to associate points that would be near, with a quasi-similar local curvature but with inverse curve orientation, we compute a last parameter that compares the normal vectors at the points P_i and P'_j. We associate to the pair of points (P_i, P'_j), the angle made by the two normal vectors :

$$\alpha(P_i, P'_j) = \widehat{\overrightarrow{c(P_i)P_i}, \overrightarrow{c(P'_j)P'_j}}$$

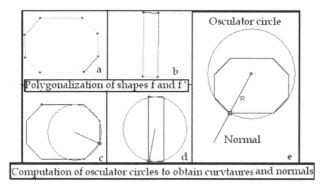

Fig. 15.1. Discrete parameters computation: polygonalization, curvature and normals. Figures a and b illustrate the polygonalization process. Figures c and d show osculator circles used in figure e to compute curvature=1/R and extract the normal as an extension of the line handling the radius.

Figure 15.1 illustrates the discrete parameters computation. We suppose, at this stage, that for each point P_i, P'_j we know its curvature, its curvilinear coordinates and its normal. We will now quantify curves matching, for this reason, lets consider a mapping M of the two curves:

$$M : [0, n] \longrightarrow [0, m], M(P_i) = P'_j.$$

Our goal is to minimize a measure of similarity on this mapping. Lets define this measure by $\phi[M] = \sum F(P_i, P'_j)$ where F is similarity cost function. F is computed by considering the distance parameter between each pair of points, the curvature difference and respective normal vectors angle. We have to precise how should F be quantified:

$$F[0, n] \times [0, m] \longrightarrow R+$$

$$F(P_i, P'_j) = d(P_i, P'_j) + \kappa(P_i, P'_j) + \alpha(P_i, P'_j)$$

Note that in order to make more robust and efficient optimization, we chose to add to coefficients k_1 and k_2 $\epsilon[0, 1]$ in order to control the influence of $d(P_i, P'_j)$ and $\kappa(P_i, P'_j)$ respectively in the optimization. By considering $k_1 = 1 - k_2$: $F(P_i, P'_j) = k_1 d(P_i, P'_j) + k_2 \kappa(P_i, P'_j) + \alpha(P_i, P'_j)$. For more details see [20].

15.3 Matching optimization

This stage is important to get correspondence for every point P_i in C with P'_j in C'.

Before proceeding with the optimization process, based on dynamic programming, lets define the structure which will be used. This structure, as shown in figure 15.2 is a grid of intersections expressing the cost $F(P_i, P'_j)$. So we can consider that our grid is $n \times m$ elements where each intersection of the axes joining P_i and P'_j is the cost of matching P_i to P'_j. This structure contains twice the same matrix for circular matching reason. When we suppose for example that P_0 is matched to P'_0, we update all other costs according to this choice. It is important to notice that the global cost function is monotic in the sense that we could not match point P_{i+1} with P'_{j-1} if we matched before P_i to P'_j as illustrated in the right part of figure 15.2. That is to say that the update for a cost between a pair of points (P_i, P'_j) could only be done by adding $min(cost(P_{i-1}, P'_{j-1}), cost(P_{i-1}, P'_j), cost(P_i, P'_{j-1}))$.

This is the application of the direct acyclic graph (DAG). The idea is to find a path which permits the matching of all chosen points of polygonalization with the minimum cost [19]. The minimum cost path determines the pairs of points to be matched while going up and picking the positions i and j.

We made tests on simple and complex shapes. For each pair of shapes, we proceed by a registration process, then we compute the intrinsic properties and finally we search the best path candidate to join all points forming the corresponding contours. We can modify the final result, by modifying the associated weighted parameters to curvature and curvilinear coordinates. The complexity is linear in the registration and parameters computing processes but logarithmic in the optimization stage. Figure 15.3 illustrates the result

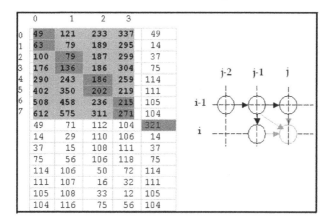

	0	1	2	3	
0	49	121	233	337	49
1	63	79	189	295	14
2	100	79	187	299	37
3	176	136	186	304	75
4	290	243	186	259	114
5	402	350	202	219	111
6	508	458	236	215	105
7	612	575	311	271	104
	49	71	112	104	321
	14	29	110	106	14
	37	15	108	111	37
	75	56	106	118	75
	114	106	50	72	114
	111	107	16	32	111
	105	108	33	12	105
	104	116	75	56	104

Fig. 15.2. Matching optimization with directed acyclic graph.

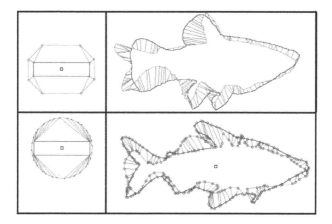

Fig. 15.3. Simple and complex shapes matching examples.

we obtain on the examples of figure 15.1 (left) and a more complex shape matching (right).

This approach as described in [20], is useful for shape recognition as we can find the most similar element of a form among others by finding the smallest cost of matching.

15.4 Model extraction

The principal idea of the point distribution model (PDM) is that each form can be defined with a set of points. This form and its model of deformation, computed over a base of the form, can be statistically studied. As we detailed in the previous sections, each form is described by a set of outline landmarks (x_i, y_i). Lets define each form from the set by $X_{1 \leq i \leq n} = (x_1, y_1, ..., x_p, y_p)^T$.

As we said before, X_i and $X_{j\neq i}$ could have different sizes (number of land-marks). This is not ideal for statistical analysis. The matching process, already describe, resolves this problem by finding for each point p_i of C_i (contour of X_i) a point p'_j in C'_j (contour of X'_j) in a context of best fitting. If we suppose that p_i is matched to two points p'_j and p'_k of C'_j, p_i should be twice added in a new vector $\hat{X}_i$ rather than X_i. Thus, we can construct new vectors $\hat{X}_i$ and $\hat{X}'_j$ with:

$$size(\hat{X}_i) = size(\hat{X}'_j) = max(size(X_i), size(X'_j))$$

This procedure can be generalized to n shapes by choosing $size(X_{1\leq i\leq n}) = max(size(X_{1\leq i\leq n}))$. All new vectors will have the same size and are ready for the statistical analysis. The linear statistical analysis consists in transforming the original data within a linear transformation optimizing a geometrical or statistical criteria. The principal components analysis (PCA) is one of the most known methods based on statistics of second order, works of Cootes and al. [2] give more details on mathematical bases. We include here these results and adapt them to our problem. PCA is a factorial method of multidimensional data analysis. It decomposes a random vector V into uncorrelated components, orthogonal and best fitting the V distribution. These components are decreasing-ordered. N representatives of X create a cloud of n points called observations. In a geometrical context, the decomposition of X is defined by the eigen vectors of the covariance matrix C:

$$C = \frac{1}{n}\sum_{i=1}^{n}(X_i - \bar{X})(X_i - \bar{X})^T \ , \ \bar{X} = \sum_{i=1}^{n}X_i$$

where $\bar{X}$ is the mean shape. C could be diagonalized to give $(\phi_i, \lambda_i)_{i=1,...,p}$, where ϕ_i are the eigen vectors and the λ_i are the eigen values. Hence, an observation X is defined by:

$$X = \bar{X} + \phi b$$

where $b = (b_1, ..., b_p)^t$ is the observation vector in the new base:

$$b = \phi^t(x - \bar{x})$$

The obtained decomposition can approximate the observations and quantify the error. We have just to select a number m, $m \leq p$, of modes in the modal base to reconstruct an observation X:

$$X = \bar{X} + \phi_m b_m$$

where ϕ_m is a sub matrix $p \times m$ of ϕ containing m eigen vectors and $b_m = (b_1, ..., b_m)^T$ is a representation in a m-dimensional space defined by the m principal components. New instances, plausible to learned observations, can be created by choosing b in the limits. Typically, the variation interval [2] is:

$$-3\sqrt{\lambda_i} \leq b_i \leq +3\sqrt{\lambda_i}$$

The new shape in figure 15.4 is created from the mean shape by combining two modes of variations.

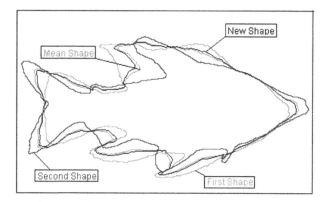

Fig. 15.4. Basic idea of the algorithm: according to mean shape and principal modes of variations compute a model able to generate plausible new shapes.

Before carrying tests, some points require to be seen. The first point is that the number of segments obtained by polygonalization can vary according to the first chosen point. At worst of cases, we will have a segment more (i.e. a point more). This is not a problem considering that the matching deals with shapes not having the same number of points on their contours(landmarks). And on another side the contour of a form can be reconstructed, with the same way, from two different polygonalizations without any loss. A second aspect of our approach is that it is very fast compared to others considering that polygonalization is useful for encoding contour of a shape, is also used to deduce the curvature in any point as well as the normal.

15.5 Results

Starting with a learning stage within a set of fishes, we proceed by the already described steps. We align shapes (scaling, rotation and translation) according to a common referential so that the sum of distances of each shape to the mean $D = \sum |X_i - \bar{X}|^2$ is minimised. The mean shape is initialized equal to the first shape and updated each time we introduce a new one. We compute then different discrete parameters after a polygonalization process applied to each shape and the mean one. We encode thus all the shapes (their landmarks) by curvatures and normals. We denote here that the curvilinear abscissa is used to course contours. Matching optimization is then done between shapes according to the mean shape. When all shapes are best matched, we compute the covariance matrix and we diagonalize it to obtain eigen values and vectors. The eigen properties describe how shapes vary linearly from the mean one. Eigen vectors give the directions of this variation and eigen values give the proportion. By analyzing different eigen values, we extract important modes of variation and we generate the model according to mean shape and the kept modes of variations. We keep always a number of modes which covers more than 90% of variations. With a set of fishes, a sample is shown in figure 15.5, we kept 24 modes of variation and by varying each mode, we obtain results in figure 15.6. If we have more shapes, it would be obvious that the number of modes will increase until reaching a limit. This is due to the fact that variations between shapes belonging to the same class are limited.

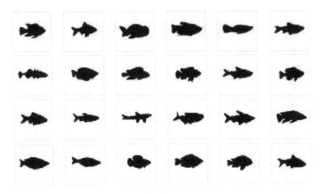

Fig. 15.5. A sample of learning data: a set of fishes.

15.6 Conclusion

We have presented a 2D shape model construction method based on shape outlines. The approach is discrete and independent of scaling and initial positions of objects. Our contribution is basically at the process of matching in opposition to methods based on manual landmarking or uniform subdivision of contours. In the proposed method, the points result from a polygonalization process that allows object reconstruction and is adapted to any 2D shape. Note that noise could be a source of inefficiency specially in polygonalization process, but this can be avoided by fuzzy polygonalization [17]. Principal components analysis allows to generate models linearly, but has also its own limits. For any particular mode of variation, the positions of landmark points can vary only along straight lines. Non-linear variation is achieved by a combination of two or more modes. This situation is not ideal, firstly because the most compact representation of shape variability is not achieved, and secondly because

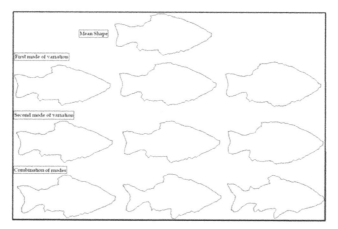

Mean Shape

First mode of variation

Second mode of variation

Combination of modes

Fig. 15.6. New statistical generated models: first row: new models according to the first mode of variation $-3\sqrt{\lambda 1} \leq b_1 \leq +3\sqrt{\lambda 1}$, the second row results from the second mode of variation and the final row is a combination of some retained modes of variation.

implausible shapes can occur, when invalid combinations of deformations are used. Attempts have been made to combat this problem. Sozou et al's Polynomial Regression PDM [16] approach allows landmark points to move along combinations of polynomial paths. Heap and Hogg's Cartesian-Polar Hybrid PDM [15] makes use of polar coordinates to model bending deformations more accurately. Sozou et al [18] have also investigated using a multi-layer perceptron to provide a non-linear mapping from shape parameters to shape. We could use these researches to improve our method. Finally, extension to 3D is under study in spite of problem of landmark order that will not be determined in a similar way between shapes.

References

1. I. Debled and J. P. Reveilles. A linear algorithm for segmentation of digital curves. Third International Workshop on Parallel image analysis, june 1994
2. T.F. Cootes and C.J.Taylor. Statistical Models of Appearance for Computer Vision. Imaging Science and Biomedical Engineering University of Manchester, March 2004, http://www.isbe.man.ac.uk
3. F. Feschet and L. Tougne. On the min DSS problem of the closed discrete curves. Discrete Applied Mathematics,151,1-3,138-153, ISSN:0166-218X,2005
4. L. Tougne. Descriptions robustes de formes en Geometrie Discrete. Habilitation a diriger des recherches de l'universite Lyon 2, december 2004
5. A. L. Yuille D. S. Cohen and P. Hallinan. Feature extraction from faces using deformable templates. International of Computer Vision, 1992,8(2), 99-112
6. L. H. Staib and J. S. Duncan. Boundary finding with parametrically deformable models. IEEE Transactions on Pattern Analysis and Machine Intelligence, 1992,14(11),1065-1075

7. M. Kass A. Witkin and D. Terzopoulos. Snakes: Active contour models. In 1st International Conference on Computer Vision, june 1987,259-268
8. C. Goodall. Procrustes methods in the statistical analysis of shape. Royal Statistical Society,1991,53(2),285-339
9. F. L. Bookstein. Principal warps: Thin-plate splines and the decomposition of deformations. IEEE Transactions on Pattern Analysis and Machine Intelligence,1989,11(6),567-585
10. J. Park D. Mataxas A. Young and L. Axel. Deformable models with parameter functions for cardiac motion analysis from tagged mri data. IEEE Transactions on Medical Imaging,1996,15,278-289
11. A. P. Pentland and S. Sclaroff. Closed-form solutions for physically based modelling and recognition. IEEE Transactions on Pattern Analysis and Machine Intelligence,1991,13(7),715-729
12. M. Turk and A. Pentland. Eigenfaces for recognition of Cognitive Neuroscience, 1991,3(1),71-86
13. M. J. Jones and T. Poggio. Multidimensional morphable models. In 6th International Conference on Computer Vision, 1998,683-688
14. B. Sebastian and P. N. Klein and B. B. Kimia and Joseph J. Crisco. Constructing 2D Curve Atlases. IEEE Workshop on Mathematical Methods in Biomedical Image Analysis,2000,70-77
15. Tony Heap and David Hogg. Extending the Point Distribution Model Using Polar Coordinates. Computer Analysis of Images and Patterns,1995,130-137
16. P.D. Sozou and T.F. Cootes and C.J. Taylor and E.C. Di-Mauro. A non-linear generalisation of PDMs using polynomial regression BMVC,1994,2,397-406
17. I. D. Rennesson and J. -L. Remy and J. Rouyer. Segmentation of discrete curves into fuzzy segments. IWCIA,may 2003,12
18. P.D. Sozou and T.F. Cootes and C.J. Taylor and E.C. Di-Mauro. Non-linear point distribution modelling using a multi-layer perceptron. BMVC,1995,1,107-116
19. Mathworld DAG. http://mathworld.wolfram.com/AcyclicDigraph.html
20. I. Boukhriss and S. Miguet and L. Tougne. Two-dimensional discrete shape matching and recogntiion. IWCIA,june 2006

A New Distorted Circle Estimator using an Active Contours Approach

Fabrice Mairesse[1], Tadeusz Sliwa[1], Yvon Voisin[1], and Stéphane Binczak[2]

[1] Université de Bourgogne, Le2i UMR CNRS 5158, Route des plaines de l'Yonne, BP 16, 89010 Auxerre Cedex, France {`Fabrice.Mairesse, Tadeusz.Sliwa, Yvon.Voisin`}`@u-bourgogne.fr`

[2] Université de Bourgogne, Le2i UMR CNRS 5158, Aile de l'ingénieur, BP 47870 21078 Dijon Cedex, France `stbinc@u-bourgogne.fr`

Summary. A new circle estimator is proposed for solving heavy subpixel error on circle center in the case of noticeable distortion. It is based on an innovative point of view over active contours process blending mathematical morphology and electronic phenomenon that furnishes a multi-level image in a single pass from the approximated form of objects to a full detailed one. A comparison with least-mean squares method supports that a sensitive improvement can be obtained compared to a more classic method.

Key words: Discrete circles, geometric noise, Radon transform, model fitting, active contours

16.1 Introduction

Active contours have been developed to overpass the limitations of classic segmentation tools as filtering, for example, leads to problems of segmentation as unconnected or incomplete contour, notably due to contrast variation, difficulty to find optimal parameters, noise, etc. Active contours are one of the best known and widely used solution to avoid these problems. The limitation of these methods is to give a unique representation of the form that can be assimilated at the scale of the representation, only allowing to have n scale representation at the price of n computation cycles. Contrariwise, mathematical morphology [1], [2] furnishes a tool that corresponds to this idea of multi-scale representation. Geodesic dilation is one of the basic mathematical morphology techniques, related to geodesic reconstruction and watershed algorithm. Although mainly numerical implementations of these techniques are available for usual real time applications [3], [4], an intuitive interpretation of these operations allows high speed imaging. A known limit of geodesic operators is that they do not take into account the regularity of the shape of the objects. But, the second connexity notion, developed by J. Serra in [2], leads to consider non-adjacent objects as being connected. In this context, the authors have decided to explore a possible generalization to the regularity of the shape of the objects. Authors have also noticed that another aspect is very important for the notion of objects' regularity: The scale from which the object is observed but multiscale analysis needs multiple wave propagation. In order to be compatible with high speed imaging, authors have studied a phenomenon producing propagation of wavefronts that have an increasing regularity along the height of the transition part. It is presented in the following section.

16.2 Active contours

Let us consider a bidimensional regular discrete (N,M) length grid Ω on which the following bistable diffusive system is defined:

$$\frac{dv_{n,m}}{dt} = D_{n,m}\left[v_{n-1,m} + v_{n+1,m} + v_{n,m-1} + v_{n,m+1} - 4v_{n,m}\right] \\ -v_{n,m}\left(a - v_{n,m}\right)\left(1 - v_{n,m}\right). \tag{16.1}$$

where $D_{n,m}$ is a local diffusion parameter and a, a threshold parameter. The system is completed by the Neumann conditions (zero-flux conditions) on the border $\partial\Omega$ of the definition domain Ω, so that:

$$\frac{\partial v_{n,m}}{\partial \eta} = 0 \ if \ (n,m) \ \in \ \partial\Omega. \tag{16.2}$$

where $\partial/\partial\eta$ denotes the outer derivative boundary.

The study of this system leads the authors to consider the emerging propagation phenomena over homogeneous and inhomogeneous grid cases to deduce a generic image processing method.

16.2.1 Homogeneous Grid Case

In order to simplify, the local diffusion parameter is considered as a constant

$$D_{n,m} = D \ \forall \ (n,m) \ \in \ \Omega. \tag{16.3}$$

This is a discrete version of the FitzHugh-Nagumo partial differential equation (PDE). In the uncoupled case, i.e. when $D = 0$, $v_{n,m} = 0$ and $v_{n,m} = 1$, $\forall\{n,m\}$ are two attracting steady states, while $v_{n,m} = a$, $\forall\{n,m\}$ is an unstable equilibrium point of the system, acting as a threshold. In case of strong coupling, i.e. when D is large, we expect that a traveling wave will propagate depending on the value of a with a constant speed so that if $a< \frac{1}{2}$ ($a> \frac{1}{2}$ resp.), the steady state $v = 1$ ($v = 0$ resp.) will propagate at the expense of the steady state $v = 0$ ($v = 1$ resp.). When $a=\frac{1}{2}$, no propagation occurs.

The response system also depends on initial conditions. For example, a marker, defined by $v_{i,j}(t=0)=1$ for a set of (i,j) values and $v_{k,l}(t=0)=0$ otherwise, imposes the kind of traveling waves due to its symmetry. The two main propagating structures are the planar and the circular wavefronts.

Planar Waves.

They emerge from a rectangular marker (Fig. 1-A) where two planar wavefronts propagate in opposing directions. Such propagation reduces the system to a one-dimensional problem that gives us:

$$\frac{dv_n}{dt} = D\left[v_{n-1} + v_{n+1} - 2v_n\right] - v_n\left(a - v_n\right)\left(1 - v_n\right) . \tag{16.4}$$

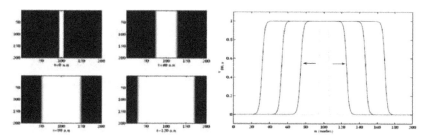

Fig. 16.1. A) A planar wavefront propagation using a grayscale representation where white corresponds to $v=1$ and black $v=0$. Top left inset shows the marker. Other insets illustrate the propagating wavefronts at different times in a.u. (arbitrary units). B) $v_{100,n}$ at different times corresponding to the insets of Fig1.a), illustrating the constant velocity. Dot line: marker, continuous lines: propagating wavefront at $t=40$, 80 and 120 a.u.. Parameters: $a=0.1$, $D=1$.

This system has been widely studied and some important results can allow us to characterize the process of propagation, although no explicit overall analytical expression of the wavefronts is available. Among these results, the differential-difference can be written in a case of strong nodes coupling, using continuum approximation:

$$\frac{\partial v}{\partial t} = \Delta v - v\left(v - a\right)\left(v - 1\right) . \tag{16.5}$$

Where Δ is the laplacian operator. A traveling-wave analysis allows us to express the traveling wave profile, considering initial conditions ξ_0 and $\xi=n-ut$, and its velocity u:

$$v\left(\xi\right) = \frac{1}{2}\left[1 \pm \tanh\left[\frac{\xi - \xi_0}{\sqrt{8aD}}\right]\right] . \tag{16.6}$$

$$u = \pm (1 - 2a) \sqrt{\frac{D}{a}} . \qquad (16.7)$$

In equation (6), one can notice the inverse relation between D and the width of the wavefront that remains true even in a discrete system. Equation (7) highlights the importance of the parameter a value compared to $1/2$ and that the velocity increases when a decreases. The $\pm$ symbol in equation (7) corresponds to the bidirectional propagation. These remarks stay qualitatively valid in the discrete case.

Moreover, a major feature due to discreteness is the failure of the propagation when D is smaller that a critical non-zero value $D^* \dot{c} 0$, which is missed in the continuum approximation. In this case, the wavefront is pinned [5, 6]. From [6], an asymptotic expression of this parameter when $a \rightarrow 0$, is

$$D^* = \frac{1}{4}a^2 . \qquad (16.8)$$

Circular Waves.

Circular waves emerge from a circular or almost circular marker or from a marker containing a circular symmetry as illustrated in figure 2.

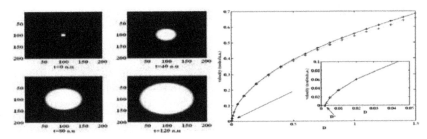

Fig. 16.2. A) A circular wavefront propagation using a greyscale representation where white corresponds to $v = 1$ and black $v = 0$. The top left inset shows the marker. The other insets illustrate the propagating wavefronts at different times in a.u. (arbitrary units). B) Propagation velocity versus the diffusive parameter D. Numerical results comparison between planar wave speed (continuous line) and circular wave speed (dot line). Parameter $a = 0.1$.

The spatial shape determines the velocity of the wavefronts, so that, in a continuous system [7], a convex traveling wave propagates slower than a planar one. This comportment is verified as it can be seen in the comparison between planar wave speed and circular wave speed when D increases (Fig. 2). When D decreases, the circular propagation is replaced by a multi-planar propagation with different wave vectors as the wavefront width becomes smaller and is confronted to the discreteness of the system. That is to say that for, a small value of D, planar and circular wavefronts are merged. In addition, propagation fails for the same value $D*$ of D. From now on, we will assume that $D*$ is not a function of the kind of traveling waves, but only determined by the threshold parameter a.

16.2.2 Inhomogeneous Grid Case

Contrary to the homogeneous grid case, the point of interest is the blocking case. Indeed, it allows the control of the path of the topological traveling wave. An image is then defined as a discrete bi-dimensional grid where nodes correspond to pixels sites. Each node is coupled to its nearest neighbors in a diffusive manner weighted by the local intrinsic information of the image. The main idea is to initiate a wavefront and let it propagate until it reaches an object Θ to be detected. In order to prevent further propagation, we now impose the following rule:

$$D_{n,m} = \begin{cases} D_p > D^* \; if \; (n,m) \notin \Theta \\ 0 \; otherwise \end{cases} . \tag{16.9}$$

I.e. the wavefront cannot propagate in the object ($D=0$) and is limited to the borders of Θ. Contrariwise, the propagation is possible since the marker (or a part of the marker) is outside the object ($Dp>D^*$) (Fig. 3-A). The relationship between the diffusive parameter D, the width W and the thickness T of the corridor has been studied and is presented in Fig. 3-C. A critical value, Dm, of the diffusive parameter leads to a no propagation of the wavefront (Fig. 3-B). Obviously, the larger and thicker the corridor is, the wider a propagating wave can be, therefore, Dm increases. As discussed in the following section, this property can be interesting to integrate objects or to develop active curves.

16.2.3 A Generic Image Processing Method

Let A and B denote sets or indicate sets functions or even grayscale image quantities as scalar-type results of some image processing (for example, linear filtering). A and B can derive from the same image or from different images. Here A represents binary marker from which a propagation phenomenon starts from, and B a topological constraint derived from the image. The couple of scalar discrete functions (A,B) defines the following equation:

$$E_\varepsilon(A,B) : \begin{cases} \frac{dv_{n,m}}{dt} = \frac{D(B)}{\varepsilon}(v_{n-1,m} + v_{n+1,m} + v_{n,m-1} + v_{n,m+1} \\ \qquad\qquad\qquad\qquad\qquad -4v_{n,m}) - f_a(v_{n,m}) \\ with \; f_a(v) = v(a-v)(1-v) \; for \; a < \frac{1}{2}, \\ D(B) = \frac{1+\tanh(20B-12)}{2}, \; and \; v|_{t=0} = A \end{cases} . \tag{16.10}$$

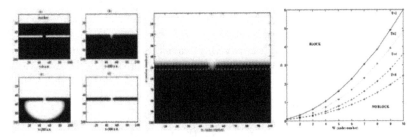

Fig. 16.3. A) Propagation of a wavefront in an inhomogeneous grid. Inset (a) shows the initial marker and an object Θ. Inset (b) shows the wave front crossing the corridor separating the two parts of Θ at $t = 100$ a.u.. Inset (c) shows the propagation of a circular wave emerging from the corridor at $t = 200$ a.u.. Inset (d) shows the final stationary state, obtained at $t = 300$ a.u.. B) Unsuccessful propagating wave (D = 1.2) C) Diffusive parameter D versus width W; curves for different values of thickness T of Θ. When D is above this curve, the wavefront is pinned. When D is beneath this curve, the wavefront can cross the corridor and propagate. Parameters: $a = 0.1$, $D = 1$, *W=6 nodes*, *T=4 nodes*.

With $a = 0.1$ and the definition domain Ω being the *(n,m)-length* grid. The system is completed by the Neumann conditions. A marker-rule is set so that the marker remains constant and equal to 1. This choice of constructing $D(B)$ corresponds to a bimodal distribution of the local diffusive parameter separated by $D*$, allowing the control of the propagating paths. The propagation phenomenon defined by E tends to a convergence state noted $v_\varepsilon^\infty(A,B)$, theoretically corresponding to infinite t but practically a millisecond value will be sufficient in most cases. Then, let us define the main image processing operator as:

$$\phi_{e,h}(A,B) \;=\; \{v_e^\infty(A,B) \geq h\}. \qquad (16.11)$$

The propagation phenomenon is similar to the geodesic propagation, but with an additive scale of regularity constraint, defined by the couple variables (ε,h) with ε defining the magnitude order of the scale of regularity of the one-pass propagation and h a thresholding along the scale regularity consequently to the choice of ε in the final result when the propagation is definitively blocked by the topological constraint. It is equivalent to fix ε and deduce h or to fix h and deduce ε. Eq. (11) leads to the immediate following properties:

When $\varepsilon \;^-\!\!\to 1+/D^*$, $\Phi_{\varepsilon,h}(A,B)$ tends to the geodesic reconstruction of Bmarked by A.

$\Phi_{\varepsilon,h}(.,B)$ is increasing with fixed regularity scale parameters and increasing with its two regularity scale parameters increasing independently.

$\Phi_{\varepsilon,h}(A,.)$ is decreasing with fixed regularity scale parameters and decreasing with its two regularity scale parameters increasing independently.

It allows the generation of many morphological-type analysis techniques, constructed as classic techniques through the morphological theory, but with some advantages from the point of view of active curves (considering the evolving front) or regions (considering the interior of the evolving curve) as their greater regularity, without the topological problems of the deformable templates. As illustrated by the experimental study, ε plays the rule of a main geometric scale parameter and h a secondary geometric scale parameter. The $>$ sign produces an active geodesic region approach, $<$ producing a dual region approach, whilst replacing it by $=$ produces an active geodesic curve approach. For the following paragraph, c represents $1/\varepsilon$.

One of the main advantages that has not been illustrated is the one-pass multi-scale approximation. Considering the shape of figure 4, the propagation of the traveling wave has a comportment of viscous gauge (Fig. 4). Analyzing the grayscale resulting image gives us a set of multi-scale representations of the original shape (Fig. 5). Low thresholds give higher details and high thresholds furnish higher approximation of the shape. At limit of the domain of the image, i.e. in our case $[0, 1]$, we obtain almost the initial shape for low threshold (~0) and approximation hulls for high threshold (~1).

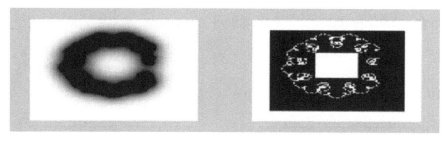

Fig. 16.4. Scale aspect: Hull effect (with inner and outer markers). Left: final state. Right: Shape and marker. Parameter c = 8.

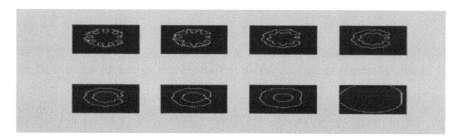

Fig. 16.5. Secondary geometrical scale aspect of h. From left to right, then top to bottom, h increases in a ratio-2 geometrical manner from h = 0.0078 to h = 0.996. Parameter: c = 8.

The figure 6 illustrates the effect of blockage depending on c. The more c increases, the less propagation penetrates in the porous medium. It works as if the propagation becomes more viscous. It could be interpreted as if each elementary part of the propagation front was blocked by a virtual dilation whose size is determined by the scale c. Therefore, c can also be denoted as the "scale of observation". This phenomenon is useful for noised circles centre determination.

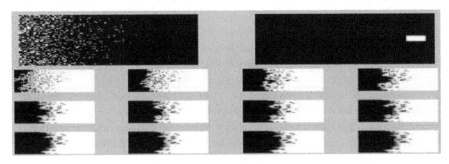

Fig. 16.6. Scale aspect of c: Porosity effect (up: shape and marker, down: propagation blockage for increasing values of c from left to right and up to bottom: c = 0.1 to c = 1.75 by incremental steps of 0.15)

16.3 Applications to Circles

16.3.1 Samples

A valuable application of these concepts is the measurement of circles, especially distorted ones. Indeed, an active contours approach of a distorted circle can lead to the obtainment of a more circular form (Fig. 7). So we consider a test set produced by modeling a certain variety of perfect and imperfect discrete circles. Circles coordinates are described by:

$$\{x = E\left[x_0 + r\cos\left(\theta\right)\right], y = E\left[y_0 + r\sin\left(\theta\right)\right]\} \ . \tag{16.12}$$

$E[x]$ being the nearest integer of x.

Noise and distortion can be added to those circles by modifying the radius:

$$\left\{\begin{array}{l} r \longrightarrow r - A_{noise} \\ r \longrightarrow r - a_{distortion} \end{array}\right\} \ . \tag{16.13}$$

with A_{noise} a random variable of centered Gaussian density of probability and $a_{distortion} = a|sin(\theta)sin(\alpha\theta)exp(-\beta\ (\theta + \pi)^4)|$, a, α and β representing parameters of this deformation.

With varying values of x_0, y_0 and r, we obtain a statistically significant set of test images to distinguish our measurement tools. Typically, noise and distortion amplitude are respectively set to 5 and 10 (Fig. 8). The relative scale unit is the pixel one.

Fig. 16.7. Left: initial image and final active contours image, Right: several thresholds of the final active contours image (i.e. several views at different scale)

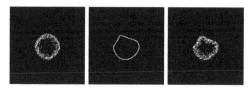

Fig. 16.8. Left: noised circle, Middle: distorted circle, Right: noised and distorted circle

16.3.2 Radon Based Method

The main idea of the method is to describe a circle by its tangents to access to circle radii. Indeed, three tangents i.e. three points are sufficient to compute circle characteristics and only two if we consider parallel tangents. The authors' approach is to compute radii from those tangents and determine circle center by their intersection. Considering a discrete framework, each arc of the circle can be considered as a segment at a certain scale. Thereby, a discrete tangent can pass through one and, generally, several pixels.

That is why the authors use the Radon transform to find tangents [8]. As the Radon transform converts an image (x,y) into a new domain (ρ, θ), it is obvious to isolate the lines which include the most of pixels because they correspond to maxima in the upper and lower parts of the signal. Although these lines do not always correspond to tangents, they are even so neighboring. In the continuation, these lines are considered as circle tangents.

The first step is to find each couple of tangents at the discrete angle θ_i ($\theta_i \in [0, \text{pi}[)$. To that end, the Radon transform is separated into two parts at the level of the barycenter for all discrete angles to obtain the upper and lower parts of the original signal. This allows to correctly identify each maximum in each part of the signal.

Once these maxima found, radii can be computed by considering the mean of the parameter ρ of the tangents for each θ_i in the Radon domain. For simplification purpose, the value of that point is set to the mean of maxima. To enhance results, a fitting is done on data considering the parametric representation of an image point, here the circle center, in Radon domain.

The authors use a trust region algorithm[9] for nonlinear least mean squares. To improve the fitting, it is necessary to suppress unlikely points that imply a loss of precision when noise or distortion is added. The implemented method is an iterative $\alpha.\sigma$ method where α decreases along iteration. In order to have the radius of the circle, a mean of the distance between each tangent is done.

16.3.3 Results

To have a point of comparison, the authors use a classic least-mean squares (LMS) method for circles [10]. Results are presented in the tables 1 and 2. Both methods have been applied on initial simulated circles, then on edge of several threshold (i.e. several scale) of the active contours results. The first conclusion is that the RB estimator is not adapted for perfect or noised circles. It only takes its interest in the case it is made for, that is to say the distorted circles. One can see that the active contours method does not significantly affected the precision of least mean square estimator. In the same way, even if an increase of precision is visible, it is only significant for distorted circles as heavy distortions are compensated. The difference of precision reaches nearly 1.5 pixel in distorted cases and nearly 0.84 pixel in noised and distorted cases in favor of RB estimator with a threshold of 128 and eight or the nine iterations.

Table 16.1. Mean error of the least mean squares method

Mean error of the least mean squares method on circle center position (given in pixel)					
	Threshold	Perfect circle	Noised circle	Distorted circle	Noised & distorte circle
Initial image		0,014744	0,13226	2,1681	2,2607
Active contours	2	0,022057	0,127	2,1448	2,0067
	4	0,022057	0,12014	2,1448	2,0286
	8	0,022057	0,11924	2,1448	2,0472
	16	0,022057	0,11253	2,1448	2,0727
	32	0,022057	0,11003	2,1448	2,0938
	64	0,025468	0,10829	2,1367	2,1188
	128	0,025539	0,11306	2,1377	2,1219

Table 16.2. Mean error of the Radon based method

Mean error of the Radon based method on circle center position (given in pixel)					
	Threshold	Perfect circle	Noised circle	Distorted circle	Noised & distorte circle
Number of iterations needed		0 iteration	2 iterations	8 iterations	9 iterations
Initial image		0,44724	0,52831	1,5025	2,8024
Active contours	2	0,40452	0,37137	0,77922	1,7336
	4	0,40452	0,36452	0,77922	1,643
	8	0,40452	0,34141	0,77922	1,6091
	16	0,40452	0,41065	0,77922	1,4555
	32	0,40452	0,33302	0,77922	1,5274
	64	0,31284	0,35953	0,69957	1,2991
	128	0,3853	0,34013	0,64396	1,1671

16.4 Conclusion

This paper has contributed a new approach for problems of circles measurement in a heavy distorted context. It consists on an active region algorithm that is able to furnish a multi-leveled image of a contour in a single pass combined with a Radon transform based estimator. Levels are defined by a granularity parameter that allows to focus on a particular scale. A more circular form can be obtained at a larger scale even if an error on the diameter is introduced. The circle estimator, resting on an approach of the circle by its tangents, furnishes a subpixel approximation of the center thanks to a fitting in the Radon parameter domain. Results, compared to various classic methods, show the adequacy of a pre-processing by a propagation of wavefronts combined with a tangential approach for measurement on distorted circles.

References

1. Serra J (1982) Image Analysis and Mathematical Morphology Vol. I. Academic Press, London
2. Serra J (1988) Image Analysis and Mathematical Morphology Vol. II : Theoretical Advances. Academic Press, London
3. Goldenberg R, Kimmel R, Rivlin E, Rudzsky M (2001) IEEE Trans. Image Proc., 10:1467-1475
4. Paragios N, Deriche R (2000) IEEE Trans. PAMI, 22:266-280
5. Keener J (1987) J. Appl. Math., 47:556-572
6. Erneux J, Nicolis G (1993) Physica D, 67:237-244
7. Fast V G, Kleber G (1997) Cardiovascular research, 33:258-271
8. Kim H, Kim J (1997) Pattern recognition letters, 22:787-798
9. Branch M A, Coleman T F, LI Y (1999) Journal of scientific computing, 21:1-23
10. Chernov N, Lesort C (2005) Journal of Mathematical Imaging and Vision, 23:239-252

Detection of Facial Feature Points Using Anthropometric Face Model

Abu Sayeed Md. Sohail and Prabir Bhattacharya

Concordia Institute for Information Systems Engineering (CIISE)
Concordia University, 1515 St. Catherine West, Montreal, Canada H3G 2W1
a_sohai@encs.concordia.ca, prabir@ciise.concordia.ca

Summary. This chapter describes an automated technique for detecting the eighteen most important facial feature points using a statistically developed anthropometric face model. Most of the important facial feature points are located just about the area of mouth, nose, eyes and eyebrows. After carefully observing the structural symmetry of human face and performing necessary anthropometric measurements, we have been able to construct a model that can be used in isolating the above mentioned facial feature regions. In the proposed model, distance between the two eye centers serves as the principal parameter of measurement for locating the centers of other facial feature regions. Hence, our method works by detecting the two eye centers in every possible situation of eyes and isolating each of the facial feature regions using the proposed anthropometric face model . Combinations of differnt image processing techniques are then applied within the localized regions for detecting the eighteen most important facial feature points. Experimental result shows that the developed system can detect the eighteen feature points successfully in 90.44% cases when applied over the test databases.

17.1 Introduction

Identification of facial feature points plays an important role in many facial image applications including human computer interaction, video surveillance, face detection, face recognition, facial expression classification, face modeling and face animation. A large number of approaches have already been attempted towards addressing this problem, but complexities added by circumstances like inter-personal variation (i.e. gender, race), intra-personal changes (i.e. pose, expression) and inconsistency of acquisition conditions (i.e. lighting, image resolution) have made the task quite difficult and challenging. All the works that have addressed the problem of facial feature point detection so far can be grouped into several categories on the basis of their inherent techniques. Geometrical shape of facial features has been adopted in several works for facial feature point localization and detection [1][2]. Each feature is demonstrated as a geometrical shape; for example, the shape of the eyeball is circle and the shape of the eyelid is ellipse. This method can detect facial features very well in neutral faces, but fails to show better performance in handling the large variation in face images occurred due to pose and expression [3]. To overcome this limitation, a variation of shape-based approaches that looks for specific shape in the image adopting deformable and non deformable template matching [4],[5], graph matching [6], snakes [7] or the Hough Transformation [8] has also been deployed. Due to the inherent difficulties of detecting facial feature points using only a single image, spatio-temporal information captured from subsequent frames of video sequence has been used in some other work for detection and tracking facial feature points [9][10]. Combination of color information from each of the facial features has been extracted and used to detect the feature points in some other works [11][12]. One of the main drawbacks of the color based algorithms is that they are applicable only to the color images and can't be used with the gray scale images. Approaches of facial feature points detection using the machine learning techniques like Principle Component Analysis [13], Neural Network [3], Genetic Algorithm [14] and Haar wavelet feature based Adaboost classifiers [15] require a large number of face images and computational time for initial training. There are also some works that have used image intensity as the most important parameter for detection and localization of facial features [16][17].

Although anthropometric measurement of face provides useful information about the location of facial features, it has rarely been used in their detection and localization. In this chapter, we have explored the approach of using a statistically developed, reusable anthropometric face model for localization of the facial feature regions as well as for detection of the eighteen most important facial feature points from these isolated regions using a hybrid image processing technique. The subsequent discussion of this chapter has been organized into the following sections: Section 2 explains the proposed anthropometric face model, Section 3 focuses on the isolation of facial feature regions using the anthropometric face model, Section 4 explains the techniques of detecting the eighteen feature points from the identified face regions, experimental results are represented in Section 5 and finally Section 6 concludes the paper.

17.2 Anthropometric Model for Facial Feature Region Localization

Anthropometry is a biological science that deals with the measurement of the human body and its different parts. Data obtained from anthropometric measurement informs a range of enterprises that depend on knowledge of the distribution of measurements across human populations. After carefully performing anthropometric measurement on 300 frontal face images taken from more than 150 subjects originated in different geographical territories, we have been able to build an anthropometric model of human face that can be used in localizing facial feature regions from face images [18]. Rather than using all the landmarks used by Farkas [19], we have used only a small subset of points and have added some new landmarks in our model. The landmark points that have been used in our proposed anthropometric face model for facial feature localization are represented in Fig. 17.1(a). It has been observed from the statistics of proportion evolved during our initial observation that, location of these points (P_3, P_4, P_6, P_7) can be obtained from the distance between the two eye centers (P_1 and P_2) using the midpoint of the eyes (P_5) as an intermediate point since distances between the pair of points (P_1 and P_3), (P_2 and P_4), (P_5 and P_6), (P_5 and P_7) maintain nearly constant proportions with the distance between the centers of the left and right eyes (P_1 and P_2). Our proposed anthropometric face model of facial feature region localization has been developed from these proportional constants (Table 17.1) using the distance between the centers of the left and right eyes (P_1 and P_2) as the principle parameter of measurement.

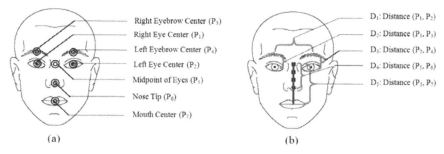

Fig. 17.1. Proposed Anthropometric Face Model for facial feature region localization (a) Landmarks used in our Anthropometric Face Model (b) Distances (Anthropometric Measurements).

17.3 Identification of the Facial Feature Regions

Since distance between the two eye centers serves as the principal parameter for measuring the center locations of other facial feature regions, implementation of the proposed automated facial feature point detection system begins with the detection of two the eye centers using the generative framework for object detection and classification proposed in [20]. This method has been able to point out the eye centers correctly almost in 99% cases for our dataset. Once the right and left eye centers (P_1 and P_2) are detected, we step forward for measuring the rotation angle of the face in a face image over the horizontal axis (x-axis). For this purpose, we have imagined a right angled triangle formed with the right eye center (x_1, y_1), left eye center (x_2, y_2) and the third point that serves as the crossing point of the horizontal and vertical lines passing through the right and left eye centers respectively (Fig. 17.2).

Table 17.1. Proportion of the distances (D_2, D_3, D_4, and D_5) to D_1 measured from the subjects of different geographical territories

Proportion	Description	Constant
D_2/D_1	Proportion of the distance between right eye center and right eyebrow center to the distance between eye centers	$\simeq 0.33$
D_3/D_1	Proportion of the distance between left eye center and left eyebrow center to the distance between eye centers	$\simeq 0.33$
D_4/D_1	Proportion of the distance between midpoint of eye centers and nose tip to the distance between eye centers	$\simeq 0.60$
D_5/D_1	Proportion of the distance between midpoint of eye centers and mouth center to the distance between eye centers	$\simeq 1.10$

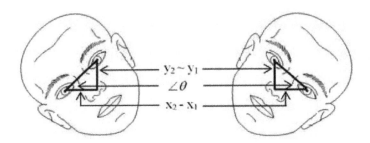

Fig. 17.2. Horizontal rotation (rotation over x-axis) of the face in a face image is corrected by determining the rotation angle. Coordinate values of the eye centers are used to calculate the amount of rotation to be performed.

The amount of horizontal rotation of the face, θ is then determined using the following equation:

$$\theta = \tan^{-1}\left(\frac{|y_1 - y_2|}{x_2 - x_1}\right)$$

For fitting the face with our model, the whole image is then rotated by the angle θ. The direction of the rotation (clock-wise or counter clock-wise) is determined by the polarity of difference (y_1 - y_2). If the difference is positive, the image is rotated in clock-wise direction, otherwise it is rotated to counter clock-wise direction. The new location of the right eye center (x_1, y_1) and left eye center (x_2, y_2) are then updated by detecting them once again over the rotated face image. Once the face is aligned to fit with our anthropometric model, the system works for detecting the centers of facial feature regions (P_3, P_4, P_6, P_7). This process begins with the detection of the midpoint of the two eye centers (P_5) that serves as the reference point for detecting P_6 and P_7, and the distance D_1 that is used as the principal parameter for measuring the location of other facial feature regions. Locations of the points

P_3, P_4, P_6, P_7 are then identified by calculating the distances D_2, D_3, D_4, and D_5 respectively Fig. 17.1(b) using the proportionality constants proposed by our anthropometric face model (Table 17.1). Rectangular bounding boxes for confining the facial feature regions are then approximated using the distance between the two eye centers as the measurement criteria.

17.4 Facial Feature Point Detection

Searching for the eighteen facial feature points is done separately within each of the areas returned by the facial feature region identifier. Steps for the searching process are described below:

17.4.1 Feature Point Detection from the Eye Region

The eye region is composed of the dark upper eyelid with eyelash, lower eyelid, pupil, bright sclera and the skin region that surrounds the eye. The most continuous and the non deformable part of the eye region is the upper eyelid, because both pupil and sclera change their shape with the various possible situations of eyes, especially when the eye is closed or partially closed. So, inner and outer corner are determined first by analyzing the shape of the upper eyelid.

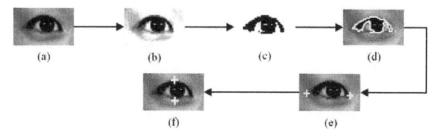

Fig. 17.3. Detection of the feature points from the eye region. (a) Eye region (b) Intensity adjustment (c) Binarized eye region (d) Detected eye contour (e) Detected inner and outer eye corners (f) Detected midpoints of the upper and lower eyelids.

To avoid the erroneous detection, discontinuity of the upper eyelid must be avoided. This can be done by changing the illumination of the upper eyelid in such a way that it differs significantly from its surrounding region. Here, this has been carried out by saturating the intensity values of all the pixels towards zero that constitutes the lower 50% of the image intensity cumulative distribution and forcing the rest of pixels to be saturated towards one Fig. 17.3(b). The adjusted image is then converted to binary one Fig. 17.3(c) using the threshold obtained from the following iterative procedure [21]:

1. Pick an initial threshold value, t.
2. Calculate the two mean intensity values (m_1 and m_2) from the histogram using the intensity values of the pixels that lie below and above the threshold t.
3. Calculate the new threshold $t_{new} = (m_1 + m_2)/2$.
4. If the threshold is stabilized ($t=t_{new}$), this is the appropriate threshold level. Otherwise, t becomes t_{new} and re-iterate from step 2.

Contour that covers the largest area Fig. 17.3(d) is then isolated using the 8-connected chain code based contour following algorithm specified in [22]. For the right eye, the inner eye corner is the rightmost point of the contour and outer eye corner is the leftmost one. For left eye, rightmost point over the contour becomes the outer corner and leftmost point is the inner corner. The whole eye contour is then divided vertically into three equal parts and searching for the upper and lower mid eyelid is then done within the middle portion. For each value of x-coordinate $\{x_1, x_2, ..., x_n\}$ that falls within this middle portion, there will be two values of y-coordinate; one from the upper portion of the eye contour $\{y_{11}, y_{12}, ..., y_{1n}\}$ and another from the lower portion of the eye contour $\{y_{21}, y_{22}, ..., y_{2n}\}$. Distance between each pair of points $\{(x_i, y_{1i}), (x_i, y_{2i})\}$ is then calculated. The maximum of the distances calculated from these two sets points and that lies closest to the midpoint of inner and outer eye corner is considered as the amount of eye opening. Mid upper eyelid and mid lower eyelid are simply the points that forms the maximum distance.

17.4.2 Eyebrow Corner Detection

Aside from the dark colored eyebrow, eyebrow image region also contains relatively bright skin portion. Since dark pixels are considered as the background in digital imaging technology, the original image is complemented to convert the eyebrow region as the foreground object and rest as the background Fig. 17.4(b). A morphological image opening operation is then performed over the complemented image with a disk shaped structuring element of ten pixel radius for obtaining the background illumination Fig. 17.4 (c). The estimated background is then subtracted from the complemented image to get a brighter eyebrow over a uniform dark background Fig. 17.4(d). Intensity of the resultant image is then adjusted on the basis of the pixels' cumulative distribution to increase the discrimination between the foreground and background Fig. 17.4(e). We have then obtained the binary version of this adjusted image Fig. 17.4(f) by thresholding it using the Otsu's method [23] and all the available contours of the binary image are detected using the 8-connected chain code based contour following algorithm specified in [22]. The eyebrow contour, which is usually the largest one, is then identified by calculating the area covered by each contour Fig. 17.4(g). For the left eyebrow, the point on the contour having the minimum values along the x and y coordinates simultaneously, is considered as the inner eyebrow corner. Again, the point which has the maximum values along both the x and y co-ordinates at the same time, is considered as the outer eyebrow corner Fig. 17.4(h). Similar, but reverse procedure is applied over the right eyebrow for detecting its inner and the outer corner.

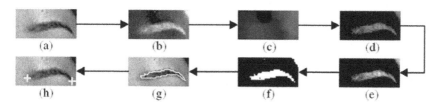

Fig. 17.4. Eyebrow corners detection (a) Eyebrow region (b) Complemented eyebrow image (c) Estimated background (d) Background Subtraction (e) Intensity adjustment (f) Binary eyebrow region (g) Eyebrow contour (h) detected eyebrow corners.

17.4.3 Detection of the Nostrils

The nostrils of the the nose region are two circular or parabolic objects having the darkest intensity level Fig. 17.6(a). For detecting the centre points of nostrils, separation of this dark part from the nose region has been performed by filtering it using Laplacian of Gaussian (LoG) as the filter. The 2-D LoG function centered on zero and with Gaussian standard deviation σ has the form [24]:

$$LoG(x,y) = -\frac{1}{\pi\sigma^4}\left[1 - \frac{x^2 + y^2}{2\sigma^2}\right]e^{-\left(\frac{x^2+y^2}{2\sigma^2}\right)}$$

Benefits of using the LoG filter is that the size of the kernel used for the purpose of filtering is usually much smaller than the image, and thus requires far fewer arithmetic operations. The kernel can also be pre-calculated in advance and so, only one convolution is needed to be performed at run-time over the image. Visualization of a 2-D Laplacian of Gaussian function centered on zero and with Gaussian standard deviation $\sigma = 2$ is provided in Fig. 17.5.

The LoG operator calculates the second spatial derivative of an image. This means that in areas where the image has a constant intensity (i.e. where the intensity gradient is zero), the LoG response will be zero [25]. In the vicinity of a change in intensity, however, the LoG response will be positive on the darker side, and negative on the lighter side. This means that at a reasonably sharp edge between two regions of uniform but different intensities, the LoG response will be zero at a long distance from the edge as well as positive just to the one side of the edge, and negative to the other side. As a result, intensity of the filtered binary image gets complemented and changes the

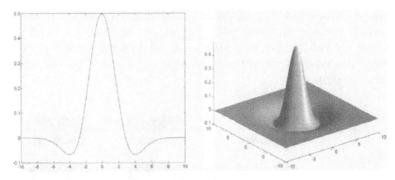

Fig. 17.5. The inverted 2-D Laplacian of Gaussian (LoG) function. The x and y axes are marked with standard deviation ($\sigma = 2$).

nostrils as the brightest part of the image Fig. 17.6(b). Searching for the local maximal peak is then performed on the filtered image to obtain the centre points of the nostrils. To make the nostril detection technique independent of the image size, the whole process is repeated varying the filter size starting from ten pixel until the number of peaks of local maxima is reduced to two Fig. 17.6(c). Midpoints of the nostrils are then calculated by averaging the coordinate values of the identified nostrils Fig. 17.6(d).

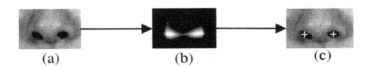

(a)   (b)   (c)

Fig. 17.6. Nostril detection from isolated nose region (a) Nose region (b) Nose region filtered by Laplacian of Gaussian filter (c) Detected nostrils.

17.4.4 Feature Point Detection from the Mouth Region

The simplest case of detecting the feature points from the mouth region occurs when it is normally closed. However, complexities are added to this process by situations like when mouth is wide open or teeth are visible between upper and lower lips due to laughter or any other expression. These two situations provides additional dark and bright region(s) respectively in the mouth contour and makes the feature point detection process quite complex. To handle these problems, contrast stretching on the basis of the cumulative distribution of the pixels is performed over the mouth image for saturating the upper half fraction of the image pixels towards higher intensity value. As a result, lips and other darker regions become darker while the skin region becomes comparatively brighter providing a clear separation boundary between the foreground and background Fig. 17.7(b). A flood-fill operation is then performed over the complemented image to fill-up the wholes of the mouth region Fig. 17.7(d). After this, the resultant image is converted to its binary version using the threshold value obtained by the iterative procedure described in [21]. All the contours are then identified applying the 8-connected chain code based contour following algorithm specified in [22] and the mouth contour is isolated as the contour having the largest physical area Fig. 17.7(f). The right mouth corner is then identified as a point over the mouth contour having the minimum x coordinate value, and the point which has the maximum x coordinate value is considered as the left mouth corner. Middle point (x_{mid}, y_{mid}) of the left and right mouth corner are then calculated and upper and lower mid points of mouth are obtained by finding the two specific points over the mouth contour which has the same x coordinate value as that of (x_{mid}, y_{mid}) but the minimum and maximum value of the y coordinate respectively.

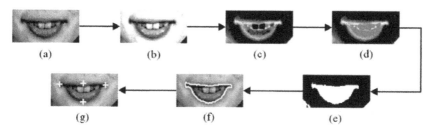

Fig. 17.7. Feature point detection from mouth region (a) isolated mouth region (b) intensity adjustment (c) complemented mouth region (d) filled image (e) binary mouth region (f) detected mouth contour (g) detected feature points from mouth region.

17.5 Experimental Results

For measuring the performance of the proposed system, we have tested it on three different publicly available face image databases namely, Caltech Face Database [26], BioID Face Database [27], and Japanese Female Facial Expression (JAFFE) Database [28]. The Caltech frontal face image database consists of 450 face images taken from 27 unique people under different lighting, expressions and backgrounds. The BioID Face Database consists of 1521 gray level images each with a resolution of 384×286 pixels. Each image in the database shows the frontal face view of one out of the 23 different test persons. The JAFFE database contains 213 images each representing seven different facial expressions (6 basic facial expressions + 1 neutral) posed by ten Japanese female models. Accuracy of our automated facial feature point detector in detecting the eighteen facial feature points over these databases has been summarized in Table 17.2. As the result indicates, our feature detector performs satisfactorily both for the neutral face images as well as for the face images having various expressions. On an average, we have been able to detect each of the eighteen feature points with a success rate of 90.44% using the proposed method.

Table 17.2. Detection Accuracy (percentage) of the Proposed Automated Facial Feature Point Detection System.

Feature Point	Caltech Face Database	BioID Face Database	JAFFE Database	Average Accuracy
Right Eyebrow Inner Corner	95.41	94.16	98.57	96.05
Right Eyebrow Outer Corner	87.36	90.42	92.17	89.98
Left Eyebrow Inner Corner	96.20	93.52	96.38	95.37
Left Eyebrow Outer Corner	88.40	86.26	90.35	88.34
Right Eye Inner Corner	93.12	90.83	94.70	92.88
Right Eye Outer Corner	85.34	87.92	89.62	87.63
Midpoint of Right Upper Eyelid	84.49	86.71	88.40	86.53
Midpoint of Right Lower Eyelid	83.60	85.38	86.73	85.24
Left Eye Inner Corner	95.11	92.64	92.83	93.53
Left Eye Outer Corner	86.69	90.76	91.46	89.64
Midpoint of Left Upper Eyelid	85.77	88.26	89.61	87.88
Midpoint of Left Lower Eyelid	84.22	87.69	88.98	86.96
Right Nostril	97.23	93.19	98.34	96.25
Left Nostril	96.95	91.88	97.21	95.35
Right Mouth Corner	92.79	87.40	95.32	91.84
Left Mouth Corner	94.10	92.45	97.89	94.81
Midpoint of Upper Lip	85.73	83.91	91.20	86.95
Midpoint of Lower Lip	79.31	82.33	86.28	82.64

17.6 Conclusion

We have presented a completely automated technique for detecting the eighteen facial feature points where localization of the facial feature regions has been performed using an statistically developed anthropometric face model [18]. One of the important features of our system is that rather than just locating each of the feature regions, it identifies the specific eighteen feature points from these regions. Our system has been tested over three different face image databases and on an average, it has been able to detect each of the eighteen facial feature points with a success rate of 90.44%. The proposed technique is independent of the scale of the face image and performs satisfactorily even at the presence of the seven basic emotional expressions. Since the system can correct the horizontal rotation of face over x-axis, it works quite accurately in case of horizontal rotation of the face. However the system has limitation in handling the vertical rotation of face over y-axis and can perform satisfactorily only when the vertical rotation is less then 25 degree. Use of the anthropometric face model [18] in the localization of facial feature regions has also reduced the computational time of the proposed method by avoiding the image processing part which is usually required for detecting facial feature regions from a face image. As the distance between the centers of two eyes serves as the principal measuring parameter for facial feature regions localization, improvement in eye center detection technique can also further improve the performance of the whole automated facial feature point detection system.

Acknowledgements. The authors wish to thank the corresponding authorities of the databases for releasing the necessary permissions regarding their use in this research. This work was supported in part by grants from the NSERC and the Canada Research Chair Foundation.

References

1. Xhang, L., Lenders, P.: Knowledge-based Eye Detection for Human Face Recognition. In: Fourth IEEE International Conference on Knowledge-Based Intelligent Engineering Systems and Allied Technologies, Vol. 1 (2000) 117-120
2. Rizon, M., Kawaguchi, T.: Automatic Eye Detection Using Intensity and Edge Information. In: Proceedings TENCON, Vol. 2 (2000) 415-420
3. Phimoltares, S., Lursinsap, C., Chamnongthai, K.: Locating Essential Facial Features Using Neural Visual Model. In: First International Conference on Machine Learning and Cybernetics (2002) 1914-1919
4. Yuille, A.L., Hallinan, P.W., Cohen, D.S.: Feature Extraction from Faces Using Deformable Templates. International Journal of Computer Vision, Vol. 8, No. 2, (1992) 99-111
5. Brunelli, R., Poggio, T.: Face Recognition: Features Versus Templates. IEEE Transactions on Pattern Analysis and Machine Intelligence, Vol. 15, No. 10 (1993) 1042-1062
6. Herpers, R., Sommer, G.: An Attentive Processing Strategy for the Analysis of Facial Features. In: Wechsler H. et al. (eds.): Face Recognition: From Theory to Applications, Springer-Verlag, Berlin Heidelberg New York (1998) 457-4687
7. Pardas, M., Losada, M.: Facial Parameter Extraction System Based on Active Contours. In: International Conference on Image Processing, Thessaloniki, (2001) 1058-1061
8. Kawaguchi, T., Hidaka, D., Rizon, M.: Detection of Eyes from Human Faces by Hough Transform and Separability Filter. In: International Conference on Image Processing, Vancouver, Canada (2000) 49-52
9. Spors, S., Rebenstein, R.: A Real-time Face Tracker for Color Video. In: IEEE International Conference on Acoustics, Speech and Signal Processing, Vol. 3 (2001) 1493-1496

10. Perez, C. A., Palma, A., Holzmann C. A., Pena, C.: Face and Eye Tracking Algorithm Based on Digital Image Processing. In: IEEE International Conference on Systems, Man and Cybernetics, Vol. 2 (2001) 1178-1183
11. Hsu, R. L., Abdel-Mottaleb, M., Jain, A. K.: Face Detection in Color Images. IEEE Transactions on Pattern Analysis and Machine Intelligence, Vol. 24, No. 5 (2002) 696-706
12. Xin, Z., Yanjun, X., Limin, D.: Locating Facial Features with Color Information. In: IEEE International Conference on Signal Processing, Vol.2 (1998) 889-892
13. Kim, H.C., Kim, D., Bang, S. Y.: A PCA Mixture Model with an Efficient Model Selection Method. In: IEEE International Joint Conference on Neural Networks, Vol. 1 (2001) 430-435
14. Lee, H. W., Kil, S. K, Han, Y., Hong, S. H.: Automatic Face and Facial Feature Detection. IEEE International Symposium on Industrial Electronics (2001) 254-259
15. Wilson, P. I., Fernandez, J.: Facial Feature Detection Using Haar Classifiers. Journal of Computing Sciences in Colleges, Vol. 21, No. 4 (2006) 127-133
16. Marini, R.: Subpixellic Eyes Detection. In: IEEE International Conference on Image Analysis and Processing (1999) 496-501
17. Chandrasekaran, V., Liu, Z. Q.: Facial Feature Detection Using Compact Vector-field Canonical Templates. In: IEEE International Conference on Systems, Man and Cybernetics, Vol. 3 (1997) 2022-2027
18. Sohail, A. S. M., Bhattacharya, P.: Localization of Facial Feature Regions Using Anthropometric Face Model. In: I International Conference on Multidisciplinary Information Sciences and Technologies, (2006)
19. Farkas, L.: Anthropometry of the Head and Face. Raven Press, New York (1994)
20. Fasel, I., Fortenberry, B., Movellan, J. R.: A Generative Framework for Realtime Object Detection and Classification. Computer Vision and Image Understanding, Vol.98 (2005) 182-210
21. Efford, N.: Digital Image Processing: A Practical Introduction Using Java. Addison-Wesley, Essex (2000)
22. Ritter, G. X., Wilson, J. N.: Handbook of Computer Vision Algorithms in Image Algebra. CRC Press, Boca Raton, USA (1996)
23. Otsu, N.: A Threshold Selection Method from Gray-Level Histograms. IEEE Transactions on Systems, Man, and Cybernetics, Vol. 9, No. 1 (1979) 62-66
24. Marr, D., and Hildreth, E.: Theory of Edge Detection. In: Royal Society of London, Vol. B 207 (1980) 187-217
25. Gonzalez, R.C., and Woods, R.E.: Digital Image Processing. 2nd edn. Prentice Hall, New Jersey (2002)
26. The Caltech Frontal Face Dataset, collected by Markus Weber, California Institute of Technology, USA; available online at: http://www.vision.caltech.edu/html-files/archive.html
27. The BioID Face Database, developed by HumanScan AG, Grundstrasse 1, CH-6060 Sarnen, Switzerland; available online at: http://www.humanscan.de/support/downloads/facedb.php
28. Lyons, J., Akamatsu, S., Kamachi, M., Gyoba, J.: Coding Facial Expressions with Gabor Wavelets. In: Third IEEE International Conference on Automatic Face and Gesture Recognition (1998) 200-205

18

Intramodal Palmprint Authentication

Munaga. V. N. K. Prasad[1], P. Manoj[1], D. Sudhir Kumar[1] and Atul Negi[2]

[1] IDRBT, Castle Hills, Road No 1, Masab Tank, Hyderabad
mvnkprasad@idrbt.ac.in,pmanoj@mtech.idrbt.ac.in
sudheerdosapati@yahoo.com
[2] Dept. of CIS, University of Hyderabad, Hyderabad
atulcs@uohyd.ernet.in

Summary. Palmprint technology is one of the biometric techniques used to identify an individual. Recognition of palmprints is based on the features like palm lines, texture, ridges etc. Several line and texture extraction techniques have already been proposed. In this paper we propose a novel technique, filiformity to extract the line like features from the palm. We also extracted texture features using Gabor filter from the palmprint image. Performance of the two techniques is determined individually. A sum rule is applied to combine the features obtained from the two techniques to develop an intramodal system. Fusion is applied at both feature level as well as matching level for the authentication mechanism. Performances of the system improved both in False Acceptance Rate (FAR) as well as Genuine Acceptance Rate (GAR) aspects in the intramodal system.

Key words: Palmprint, Filiformity, Gabor filter, FAR, GAR, Intramodal

18.1 Introduction

Biometrics is considered to be one of the robust, reliable, efficient, user-friendly secure mechanisms in the present automated world. Biometrics can provide security to a wide variety of applications including access to buildings, computer systems, ATMs [1]. Fingerprints, Iris, Voice, Face, and Palmprints are some of the different physiological characteristics used for identifying an individual. A palm is the inner surface of the hand between the wrist and the fingers [2]. Some of the features extracted from palms are principal lines, wrinkles, ridges, singular points, texture and minutiae. Principal lines are the darker or more prominent lines present on the palm. In general, three principal lines are found on palms of most individuals, namely, heartline, headline and lifeline. Wrinkles are the thinner lines concentrated all over the palm. A palmprint image with principal lines and wrinkles represented is shown in Fig 1. However there are some difficulties and issues in extracting the line features from palmprints [18-19]: 1. Line features cannot be extracted satisfactorily from images of low-resolution. 2. When we apply edge detection algorithms, fixing an edge threshold is a major problem. Choosing a low threshold ensures that we capture the weak yet meaningful edges in the image, but it may also result in an excessive number of 'false positives'. Too high a threshold, on the other hand, will lead to excessive fragmentation of the chains of pixels that represent significant contours in the image. So unless the threshold is fixed properly lines will not be extracted accurately and an exact threshold needs to be found from experimentation. 3. Lines on palms have varying widths. Unless the filter size is set correctly lines will not be extracted with their original width. 4. In case of gray level images, a line is generally darker than the background on which it is drawn. Certain parts of lines can cross-zones that are darker than they are. Extraction of such parts of a line is difficult.

In the literature different line extraction techniques were proposed, which includes edge detection techniques and line tracing techniques. Wu et al [2] used directional line detectors to extract the principal lines. Chih -Lung et al. [9] used Sobel edge detection and performed directional decomposition. Han et al., [4] used morphological operations to enhance the palm lines and employed the magnitude of the image to compute the line like features. Wu. et. al.[5] performed morphological operations to extract the palm lines explicitly and employed coarse to fine level strategy. Algorithms such as the stack filter [10] are able to extract the principal lines. Principal lines provide a commonly accessible or intuitive means to match palms. However, these principal lines by themselves may not be sufficient to uniquely identify an individual through their palmprints because different people may have similar principal lines on their palmprints [7]. To further identify a subject more precisely additional modalities are required. Ming-Cheung et. al. [17] proposed intramodel and intermodel fusion for audio visual biometric authentication. Obviously obtaining multi-modal biometric information about individuals is a difficult proposition due to the fears of identity theft and the need to keep such information secure. Therefore while multi-modal fusion might help in improving specificity of recognition it may not always be advantageous. This leaves us with the other alternative that is, intramodal fusion where the scores of multiple samples obtained from the same modality are combined to improve the effectiveness of a biometric.

In this paper we bring in two methods for the palm biometric, and the methods are extraction of lines features and that of extraction of palm texture features. Here we find a novel application of the filiformity [6] concept. Filiformity technique can be used to extract lines even from low contrast images. To extract texture features we employ the Gabor filter technique [7]. Further the information from these are fused using two different fusion strategies. The rest of the paper is organized as follows. Section 2 deals with the image acquisition, preprocessing and segmentation. Section 3 deals with filiformity technique of line extraction. Texture feature extraction is discussed in Section 4. Section 5 presents the feature extraction technique and Section 6 determines the matching technique used. Experimental results are provided

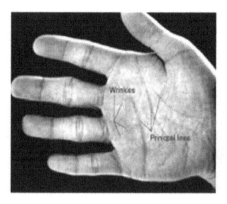

Fig. 18.1. Palmprint with principle lines and wrinkles

in Section 7. Information fusion is discussed in Section 8 and Section 9 gives the conclusions.

18.2 Image acquisition, Pre Processing and Segmentation

Any biometric recognition system has to undergo four stages like Image Acquisition, Pre-Processing and Segmentation, Feature extraction and Matching. The basic setup of the verification system is depicted in Fig 2. Palmprints of an individual can be captured in different ways like inked image [3], scanned images with pegged setup [7] and scanned images with peg free setup [4]. In our paper we worked on scanned images with pegged setup as shown in Fig 3.

Fig. 18.2. Palmprint verification system block diagram

The palmprint obtained in image acquisition is processed to smoothen and reduce the noise in the image. Adaptive median filter [8] is used to smoothen the image before extracting the ROI and its features. The filter has the advantages of handling impulse noises of large spatial densities. An additional benefit of the adaptive median filter is that it seeks to preserve detail while smoothing non-impulse noise, something that the traditional filter does not do. The images captured are to be segmented to extract the Region of Interest (ROI). The image is rotated 90 in clockwise direction as shown in Fig

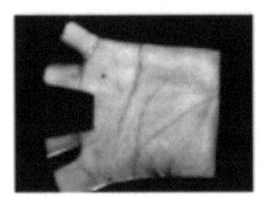

Fig. 18.3. Palmprint image captured

4. We followed the technique used by [9] to extract the square ROI from the palmprint image captured. The starting point of the bottom line Ps is found out by scanning the image from the bottom left most pixel. The boundary pixels of the palm are traced and collected in a vector, namely Border Pixel Vector as shown in Fig 4(a). The mid point of the bottom line Wm is found out and a distance to all the border pixels from Wm is calculated. A distance distribution diagram is plotted as shown in Fig 4(b). The local/minima in the distance distribution diagram are found out, which are nothing but the finger web locations. A square region is extracted using the finger web locations as shown in Fig 4(c). As the size of the square region differs from person to person, we resized the ROI to 150x150 size

18.3 Filiformity

Salim Djeziri et al., [6] proposed a method that is based on the human visual perception, defining a topological criteria specific to hand written lines which is called 'filiformity'. This method was proposed for the extraction of signatures from bank check images. Filiformity is found from two step process. The first step computes a response image where linear objects are detected. A global threshold is applied on the response image to obtain the globally relevant image objects. As far as our problem is concerned, we are only interested in the lines present on the palmprint image. A line from filiformity concepts is characterized by means of the following heuristic: A line is generally darker than the background on which it is drawn. In our approach we use ring level filiformity [6] because it is richer in terms of information and enables better interpretation of the local shape of a line by an adequate choice of a measurement function, while the surface measure translates an occupation rate without taking into account the local shape of the line. We do not perform global processing after the local values obtained because we are interested

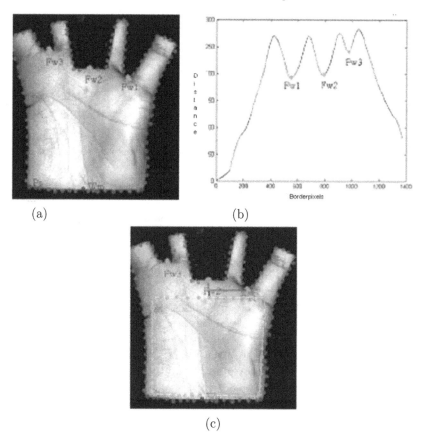

Fig. 18.4. (a) Border pixels collected and Wm point is shown (b) Distance Distribution diagram plotted for the distance between Wm and border pixels (c) Finger Web locations Fw1, Fw2, Fw3 are found and a square region is extracted.

in all the lines present on the palm. Global Processing is performed to extract the lines whose pixels have the local measure more than the specified preset threshold value. For every pixel on the palmprint image a degree of perceptibility is computed. This measure is used to build the feature vector. Filiformity technique is applied on the 150X150 size ROI and local measures are obtained for each pixel. Fig 5 shows the ROI after the application of filiformity technique.

18.4 Gabor filters for texture extraction

Palmprint have unique features like principal lines, wrinkles, and ridges. Principal lines extracted from the palmprint may not be sufficient to represent uniqueness of the individual, because of the similarity of the principal lines

Fig. 18.5. ROI after the application of filiformity

of different person's palms [7]. So to improve the uniqueness and make the feature vector much robust, we tried to extract the texture features from the images. These texture features not only include principal lines but also include wrinkles, ridges. We have used Gabor filter to extract the texture features. These Gabor filters already have been used to extract the features from fingerprint, iris recognition [11], and palmprint recognition [7] [12]. Gabor filters extracts the texture features by capturing the frequency and orientation information from the images. The 2-D Gabor filters used for palmprint verification in spatial coordinates are defined as

$$G(x, y, \theta, u, \sigma) = (\frac{1}{2\Pi\sigma^2}) * exp(-\frac{x^2 + y^2}{2\Pi\sigma^2}) * exp(2 * \pi * i * u(x \cos\theta + y \sin\theta))$$
(18.1)

Where 'x' and 'y' are the coordinates of the filter, 'u' is the frequency of the sinusoidal wave, 'σ' is the Gaussian envelope, 'θ' is the orientation of the function and i= $\sqrt{-1}$. Here, the optimized values for the Gabor filter parameters such as u=0.0925 and σ=7.1 have chosen after testing with different values at an orientation of 45.

$$I_\theta(i, j) = \sum_{x=1}^{w} \sum_{y=1}^{w} G_\theta(x, y) I(i - x, j - y)$$
(18.2)

where I is the input image, I_θ is the image at θ orientation, and w x w is the size of the Gabor filter mask. Gabor filter is applied on 150x150 size ROI as done in filiformity technique.

18.5 Building Feature Vectors from Palms

After extracting the reliable features from the palmprint image, we need to build a feature vector to represent the individual. In our approach we divided the extracted ROI into 'n' non-overlapping blocks and calculated the standard deviation of the local values obtained in both the cases of line and texture extraction. A feature vector of size 1x n is established with the computed standard deviation values. FV=[SD (1), SD (2), .SD (n)] where FV is the feature vector SD (j) is the standard deviation of the jth block.

18.6 Matching

Matching algorithm determines the similarity between two given data sets. Applying the matching algorithm on the input palmprint image and image existing in the database does palmprint verification. The palmprint is said to be authentic if the result obtained after matching is more than the preset threshold value. In our approach we employed Pearson Correlation Coefficient to find the similarity between two palmprint images. The linear or Pearson correlation coefficient is the most widely used measurement of association between two vectors. Let x and y be n-component vectors for which we want to calculate the degree of association. For pairs of quantities (xi, yi), i=1,,n the linear correlation coefficient r is given by the formula:

$$r = \frac{\sum_{i=1}^{n}(x_i - \overline{x})(y_i - \overline{y})}{\sqrt{\sum_{i=1}^{n}(x_i - \overline{x})^2}\sqrt{\sum_{i=1}^{n}(y_i - \overline{y})^2}} \qquad (18.3)$$

where $\overline{x}$ is the mean of the vector x, $\overline{y}$ is the mean of the vector y. The value r lies between -1 and 1, inclusive, with 1 meaning that the two series are identical, 0 meaning they are completely independent, and -1 meaning they are perfect opposites. The correlation coefficient is invariant under scalar transformation of the data (adding, subtracting or multiplying the vectors with a constant factor).

18.7 Experimental Results

We experimented our approach on Hong Kong PolyTechnic University Palm-print database [13]. A total of 600 images are collected from 100 different persons. Images are of 384x284 size taken at 75dpi resolution. The images, which are segmented to ROI, are resized to 150x150 size so that all the palm-prints are of same size. We applied filiformity as well as Gabor filters on the ROI independently. The resultant image is then segmented into 36 non-overlapping square blocks as shown in Fig.6. Standard deviation of the values obtained is computed for each block and a feature vector of size 1x36 is built for both the cases. The feature vectors for the six images of a person are stored in the database for his identity. FAR and GAR rates are calculated for both the cases to evaluate the performance of the system [14]. A table is built for each person in the database as shown in Table1 and Table 2. Table1 represents the correlation values of the different images that belong to the same person, which is calculated using filiformity method. Table 2 represents the correlation values of the different images that belong to the same person and is calculated using Gabor filters. The database is tested for different threshold values to calculate the GAR. A person is said to be genuine if at least one value in the table is above the threshold. FAR is calculated by matching each image of a person with all the images of the remaining persons in the database and the results are given for different threshold values. Table 3 and Table 4 presents the GAR and FAR at different threshold values for Filiformity and Gabor filter techniques respectively. Correlation between same images always yields a matching score of one. Using filiformity technique the FAR rates are

Fig. 18.6. ROI divided into 36 blocks

	2	3	4	5	6
1	0.8792	0.8189	0.8412	0.8173	0.8485
2		0.8672	0.8277	0.8910	0.8007
3			0.8070	0.7543	0.7663
4				0.8583	0.9187
5					0.9028

Table 18.1. Matching scores for different images of a person computed using Filiformity technique

better than GAR rates, which implies that a genuine person may be denied but an impostor is not allowed. In case of Gabor filter technique GAR rates are better than FAR rates, which implies that a genuine person should not be denied even though the impostor can be allowed.

	2	3	4	5	6
1	0.7202	0.7557	0.6333	0.6957	0.6152
2		0.6362	0.4547	0.6186	0.4894
3			0.7418	0.7757	0.8193
4				0.7623	0.7986
5					0.8418

Table 18.2. Matching scores for different images of a person computed using Gabor filter

Threshold	GAR	FAR
0.84	100%	6.17
0.88	99%	1.54
0.89	98%	0.96
0.9	98%	0.57
0.92	93%	0.12

Table 18.3. GAR and FAR rates for different threshold values for Filiformity technique

Threshold	GAR	FAR
0.84	100%	0.22
0.88	96%	0.024
0.89	95%	0.011
0.9	89%	0.0033
0.92	74%	0

Table 18.4. GAR and FAR rates for different threshold values for Gabor filter technique

18.8 Information Fusion

Fusion of the palmprint representations improves the performance of the verification system [15]. The representations can be combined at different levels: at feature level, at score level, and at decision level. In this paper we combined the representations using a sum rule at both feature level and score level as shown in Fig.7 and Fig.8 respectively. According to the experiments conducted by Arun Ross et al., [16] the sum rule performs better than the decision tree and linear discriminant analysis. FAR and GAR rates are calculated from the feature vectors obtained using fusion at feature level and the results are shown in Table 5. We also performed fusion at matching level by applying sum rule on the matching scores obtained from the two different techniques and calculated GAR and FAR. Results are shown in Table 6. By combining, the two techniques using fusion strategies the performance is improved in both GAR as well as FAR cases.

Threshold	GAR (%)	FAR(%)
0.8	100	0.318
0.81	98	0.218
0.82	96	0.142
0.83	96	0.085
0.84	95	0.050

Table 18.5. GAR and FAR rates at feature level fusion

Threshold	GAR (%)	FAR(%)
1.68	100	0.32
1.76	97	0.25
1.78	97	0.249
1.8	96	0.248
1.84	88	0.248

Table 18.6. GAR and FAR rates at matching level fusion

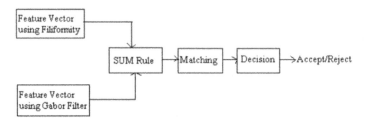

Fig. 18.7. Block diagram for feature level fusion

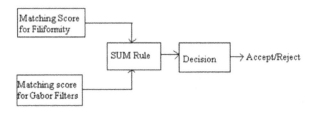

Fig. 18.8. Block Diagram for score level fusion

18.9 Conclusions

The objective of this work is to investigate the performance of the intramodal palmprint system by extracting reliable features from the palmprint. We extracted line like features as well as texture features from the palmprint image. Performance rates like GAR and FAR are calculated for each technique. In case of filiformity technique the FAR rates are better than GAR rates, which implies that a genuine person may be denied but an impostor is not allowed.

In case of Gabor filter technique GAR rates are better than FAR rates, which implies that a genuine person should not be denied even though the impostor can be allowed. By combining, the two techniques using fusion strategies the performance is improved in both GAR as well as FAR cases. While GAR is increasing FAR is decreasing by the application of fusion strategy. This concludes that using multiple features for a single biometric yields better results when compared to using single feature.

References

1. S. Nanavati,Michael Thieme, and Raj Nanavati eds.(2002) Biometrics: Identity Verification in a Networked World. John Wiley and Sons.
2. X. Wu, David Zhang, K. Wang, Bo Huang (1999) " Palmprint classification using principal lines", Pattern Recognition 37 : 1987- 1998.
3. D. Zhang, W. Shu (1999) "Two novel characteristics in palmprint verification: datum point invariance and line feature matching", Pattern Recognition 32 : 691-702
4. C.C. Han, H. L. Chen, C.L. Lin, K. C. Fan (2003) "Personal authentication using palmprint features, Pattern Recognition 36 : 371-381.
5. X. Wu, K. Wang (2004) "A Novel Approach of Palm-line Extraction", Proceedings of the Third International Conference on Image and Graphics (ICIG'04) pp 230-233.
6. Plamondon R, Djeziri S , Nouboud F(1998) IEEE Transactions on Image Processing 10: 1425-1438.
7. D. Zhang, Wai-Kin Kong, Jane You and M.Wong (2003) "Online palmprint Identification", IEEE Trans. On Pattern Analysis and Machine Intelligence, Vol.25, No.9 : 1041-1050.
8. Rafael C. Gonzalez and Richard E. Woods eds(2003) Digital Image Processing, Low Price Edition 2nd edition.
9. Chih-Lung Lin, Thomas C. Chuang, Kuo-Chin Fan (2005)" Palmprint Verification using hierarchical decomposition", Pattern Recognition 38, No. 12 : 2639-2652.
10. Paul S. Wu, Ming Li (1997) "Pyramid edge based on stack filter", Pattern Recognition 18 : 237-248.
11. John G. Daugman (1993) " High Confidence Visual Recognition of Persons by a Test of Statistical Independence" IEEE Trans. On Pattern Analysis and Machine Intelligence, Vol.15, No.11 : 1148-1161.
12. Wai Kin Kong, David Zang (2002) "Palmprint Texture Analysis based on low resolution images on personal authentication", ICPR 2002,pp 807-810.
13. "http://www.comp.polyu.edu.hk/ biometrics", PolyU Palmprint Database, Hong Kong PolyTechnic University
14. M.Golfarelli, Daraio Maio, D. Maltoni (1997) " On the Error-Reject Trade-Off in Biometric Verification Systems", IEEE Trans. On Pattern Analysis and Machine Intelligence, Vol.19, No.7 : 786-796.
15. Ajay Kumar, D. Zhang (2005) " Personal authentication using multiple palmprint representations", Pattern Recognition 38, No. 10 : 1695-1704.
16. Arun Ross, Anil Jain (2003)" Information fusion in biometrics", Pattern Recognition Letters 24:2115-2125.

17. Ming-Cheung Cheung, Man-Wai mak and Sun-Yuan Kuan (2004) "Intramodal and intermodal fusion for audio-visual biometric authentication", International symposium on Intelligent multimedia, video and speech processing, Hong Kong pp:25-28.

18. J. Canny (1986) "A computational approach to edge detection", IEEE Trans. Pattern Analysis and Machine Intelligence Vol. 8, No. 6:79-698.

19. Rafael C. Gonzalez and Richard E. Woods Eds(2003). "Digital Image Processing", Pearson Publication 2nd edition.

Part IV

Multimedia Processing

An Implementation of Multiple Region-Of-Interest Models in H.264/AVC

Sebastiaan Van Leuven[1], Kris Van Schevensteen[1], Tim Dams[1], and Peter Schelkens[2]

[1] University College of Antwerp
   Paardenmarkt 92, B-2000, Antwerp, Belgium,
   t.dams@ha.be
[2] Vrije Universiteit Brussel
   Pleinlaan 2, B-1050, Brussel, Belgium,
   peter.schelkens@vub.ac.be

19.1 Introduction

In our fast evolving time, where society, information and communication are getting closer together, the need for mobile communications is growing quickyl. Mobile phones are no longer solely used for making telephone calls. They can now take pictures, capture video, check e-mail and let the user browse the internet. Most of these applications are pretty common nowadays; on the video part however there's still a long road ahead. The mobile operators are expanding their networks to make high-speed transmissions of such data possible, however for the time being bandwidth remains a scarce good.

Because most networks deal with a limited amount of bandwidth, scaling techniques are introduced to send less data over the network with as little inconvenience as possible for the user. One of these techniques is Region-Of-Interest coding (ROI). ROI will divide an image into multiple parts, the most important part being the one the user is observing, called the ROI.

The background will be sent in a lower quality than the ROI, or not sent at all. This results in a lower bitrate and thus less bandwidth is required for the encoded video. The ROI can be defined by the user by means of a mouse click, by making use of an eye tracking device or can be predicted, based on content recognition algorithms. In this chapter, we will discuss early ROI implementations in H.264 and suggest our own solution for some of the problems discovered previously. Firstly however, we will discuss the features of H.264 that enable ROI coding.

The remainder of this chapter will have the following outline: Section II discusses the basic concepts related to H.264 and ROI coding. Section III describes our implementation of the ROI model. In section IV, we summarize the tests performed. Finally, section V provides a summary.

19.2 Basic Concepts

19.2.1 H.264

The H.264 video codec, also known as MPEG-4 Part 10 /AVC (Advanced Video Coding), has recently been standardized by MPEG [5] [6]. The original H.264 standard includes 3 profiles (baseline, main and extended), each having a different set of functionalities. The baseline profile is our main focus for this research, since it was designed primarily for low-cost applications which do not have a great amount of computational power. This profile is mainly used in mobile applications. The most relevant functionalities in the baseline profile are, Flexible Macroblock Ordering (FMO) and Arbitrary Slice Ordering (ASO). Both techniques are used for manipulating the decoding order of the macroblocks in the picture. This was implemented as an error robustness feature, but FMO and ASO can also be used for other purposes such as Region-Of-Interest.

Slices are a very important improvement in H.264. A video frame consists of macroblocks which can be grouped into slices. A slice contains at least one macroblock but can be extended to all the macroblocks in the video frame. For the Baseline Profile only I (Intra)- and P (Predicted) slices can be used and therefore only I- and P-macroblocks are supported (B-(Bidirectional), SI-(Switching) and SP- (Switching P-)slices are not supported in the Baseline Profile).

A slice group contains macroblocks of one or more slices. The macroblocks in a slice group are coded sequentially. The standard includes 7 modes to map macroblocks to a slice group[6], but we only review the 3 types relevant for our research:

- **Foreground and background (Type 2)** allows multiple rectangular slice groups to be created. When all these slice groups are filled with the corresponding macroblocks, the remaining macroblocks will be put in the last slice group. Slice groups are defined by assigning the two parameters, TOP_LEFT and BOTTOM_RIGHT, to the Macroblock (MB) numbers. When two rectangles overlap, the slice group with the lowest number will have priority and overlapping MB's will be assigned to the lowest slice group. A slice group can not exist in the following cases:
    - One of the two defining parameters is not specified;
    - BOTTOM_RIGHT is left or above TOP_LEFT;
    - The slice group is completely covered by lower slice groups;
- **Box-out (Type 3)** means 'slice group 0' starts in the center of the screen and expands spirally as a square with a predefined size. All the macro-

blocks, not contained by the box, are part of 'slice group 1'. The direction of the box-out can be either clockwise or counterclockwise;

- **Explicit (Type 6)** allows the user to define a slice group to each of the macroblocks independently. Initially this will be done by a configuration file.

The other types (such as raster scan, wipe out, check board pattern, etc.) were not used because they do not allow us to create an actual shape.

19.2.2 Region-Of-Interest

Region-of-interest coding can be used to encode objects of interest with a higher quality. The remainder of the image can be regarded as background information and can thus be encoded more coarsely. The advantage of this method is that the image parts that the viewer is looking at can be transmitted with a higher quality.This technique can be combined with other techniques such as progressive coding. The result is that the overall viewing experience remains highly satisfactory, while the transmission can be performed at lower bitrates.

Another advantage of ROI-coding is that the ROIs can be transmitted first. This can be realized by the use of slices (e.g. if 'slice group 0' is transmitted first, by placing the ROI in 'slice group 0', it should arrive first at decoder side). When network congestion occurs, the probability of having a frame that contains at least something the viewer most likely wants to see, is higher with ROI coded imagery than without ROI. Nevertheless, when a transmission error occurs at a header, inside the ROI data or in the packet length indicator, the stream can not be decoded at all. This can partially be prevented when a small ROI is chosen, which makes the probablility of an error inside the ROI lower.

There are different models to implement ROI-coding in H.264 (this is discussed in section III). These models share one trait: they all make use of slice groups. When using I-slices, the ROI will always be visible where it is defined by the user. This is because no references are needed for the encoding of I-slices. On the other hand, when P-slices are used and the position of the ROI changes, but without changes in the previous ROI (by means of motion vectors, e.g., static background) both the old and new ROI will still be visible. This is because for the old ROI no additional information is sent and the old data still will be used. This creates the possibility for having two visually more attractive regions. The current ROI ('slice group 0') and the previous ROI contain the same encoded data because no additional data is sent for these macroblocks.

19.3 Implementation

Our implementation is targeted towards an application which uses interactive, user-defined ROIs (e.g. by means of a mouse pointer). Similar research about ROI in AVC already supported the efficency of ROI coding [4] [3]. These models both use FMO type 2. Our research was focussed on the usefullness of the other FMO types. The ROI will be defined by a variable called ROIPOS, which contains the MB number of the central ROI position (i.e. the exact mouse position). An algorithm will determine which of the surrounding MB's are contained in the ROI.

The model's implementation contains 3 major parts: ROI creation, ROI check and xROI.

1. ROI Creation

During the ROI creation, different FMO types are distinguished. Each one of them needs a different approach to update the ROI.

- **FMO Type 2.**
  Both coordinates, respectively TOP_LEFT and BOTTOM_RIGHT, are computed at the encoder side;
- **FMO Type 3.**
  This model allows a box-out kind of shape. For ROI, this is restricted to a square. In order to create the square, the number of MBs needed is calculated depending on the size of the ROI. After this, the number of MBs is assigned to the parameter controlling the box-out (i.e. SLICE_GROUP_CHANGE_RATE_MINUS_1);
- **FMO Type 6.**
  For this type, every MB is checked to find out if it is part of the ROI (depending on the size and selected shape). Four shapes are considered in this model (Square, Rectangle, Diamond and Octagon). Firstly, to be able to use the square shape for all the models, so they can be compared. Secondly, to find out if it would be useful to use other shapes. At last, we wanted to implement different shapes to prove one is not constrained to symmetric structures, but amorph structures can also be implemented. Figures 19.1, 19.2 and 19.3 show the FMO map for 3 shapes (ROI=0, xROI=1, non-ROI=2). 'Slice group 0' contains the ROI for the selected shape.

An amorph (x)ROI could be interesting when one would choose to encode specific structures contained in the image, or when putting all moving MBs in one slice group by using an activity map [3]. This way not only FMO type 2 would be used to break up the screen in different rectangles, but also FMO type 6 could contain those macroblocks, having a higher Mean Absolute Difference (MAD) than a certain threshold. An algorithm then of course needs to be designed to determine the ROI. To do this, each macroblock should be checked separately to find out if it is part of the ROI.

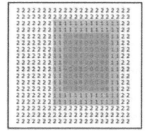

Fig. 19.1. Square

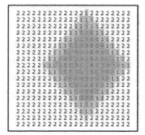

Fig. 19.2. Possible ROI Shapes. ROI=0, xROI=1, non-ROI=2

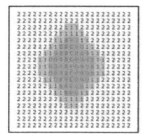

Fig. 19.3. Possible ROI Shapes. ROI=0, xROI=1, non-ROI=2

2. ROI Check

Apart from the multiple shapes, a variable position has to be supported, except for FMO type 3. This position update is sent from the decoder side. We propose to let the decoder use UDP packets to send the MB number which corresponds with the ROI. The decoder will only send this position after a change of position has occurred. After checking the value, the new ROI will be checked to know if it will keep its size. This is done to make sure the viewer's attention is not distracted by the varying size of the ROI, which can occur when the ROI is located at a border of the image. This means that, when pointing to a MB which is one of the first or last MBs of a row, the RoiPos will be moved respectively to the right or the left. The same will happen for the upper and lower border of the screen.

3. xROI

A third improvement over the other models is the extension we added to the ROI, called extensible ROI (xROI). The xROI smoothes the disturbing edge between the ROI and the non-ROI areas. Typically when a large difference exists between both Quantisation Parameters (QP), a visually disturbing transition between the ROI and non-ROI-coded image part is present. By using the proposed xROI, a controllable option is introduced, to apply an intermediate QP, in between the ROI QP and the non-ROI QP. Apart from that, another parameter, xROI, will allow the user to disable xROI completely if desired. Figures 19.1, 19.2 and 19.3 show an extensible Region-Of-Interest with width 2, applied to 3 different shapes.

Issues concerning border checking have been accounted for. When the ROI is applied close to the border of the frame (after checking or 'native'), xROI is not applied for the bordering of the ROI. And thus, no additional replacement of the RoiPos needs to be done. Figure 19.4 shows the used FMO map and the resulting image figure 19.5 ; xROI is disabled on the right side, while only a part of the bottom xROI is maintained.

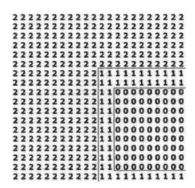

Fig. 19.4. FMO map

The xROI will have lower priority than ROI but will have higher priority than non-ROI parts of the picture. This results in slice group 1, which will contain all xROI MBs. For FMO type 3 xROI can't be used because FMO type 3 can only contain 2 slice groups: "The box" and the rest of the screen. For FMO type 2 the two coordinates are calculated and in case of the picture's border, the parameters are adjusted in such a way that it fits the previously described border checking methodology. In case of FMO type 6, an additional check is included while checking the ROI, to see if each MB makes part of the xROI.

Once the slice groups are created, it is just a matter of applying the correct quantization parameter settings to the slice groups. This quantization parameter will be applied to all MBs contained in the slice group.

Fig. 19.5. Resulting image

19.4 Test Cases

In our experiments we compared multiple ROI models. Comparisons between the usage of ROI and non-ROI have been published in [4] [3]. Implementations and tests were based on the JM10.2 reference software [1].

19.4.1 FMO type Comparison

We generated streams for FMO types 2 and 6, both static and dynamic ROI, with or without xROI, for the square and rectangle shapes. Additionally, a stream was encoded using only a static FMO type 3 square shaped ROI (moving ROI's are not supported by FMO type 3).

All the simulations used the same QP settings and RoiPos. This was done to ensure that the encoding speeds are equal, except for the slice group updating mechanisms, and to make sure any change in bitrate was due to the updates of the picture parameter set. A parameter set consists of 3 QPs (e.g. 25-30-35), first is the QP for the ROI-area (25), second the QP for the xROI-area (30) and last the QP for the background (non-ROI area (35)). For this test we used (25-30-35) as QP set.

The test sequence "Football" (4:2:0, CIF, 30 fps) was used and the RoiPos was set to 209. This position corresponds with the MB number where box-out starts counting and thus all the static ROI tests have the same visual characteristics. The results are shown in Table. 19.1. It can be clearly seen that the difference in terms of bitrate between the different FMO types is not large, while the PSNR (compared to the original test sequence) remains largly the same. The difference in bitrates is mainly due to the difference in picture parameter set information which has to be sent. For Type 2 and 6 the difference in bitrates is due to the different picture parameter set information that has to be sent. Both types send the same data but will have a different type of signalisation. Type 3 can have another bitrate due to the fact that the MB's are ordered differently such that entropy encoding can cost (or save) some bandwidth. It should be noted that the encoding time is expressed in seconds in Table 19.1 but is based on non-optimized code. The tests were done on a single threaded machine, Laptop, 1.4GHz AMD Mobile under MS Windowswith no other tasks running but the standard services. Altough a

Table 19.1. Comparison of the different FMO types

FMO TYPE comparison		Map type	Bitrate (kbps)	PSNR (Y)	Encoding time	Bits for parameter sets
ROI	Dynamic	2	1180,58	32,88	107,114	628
		6	1181,78	32,88	106,058	6884
	Static	3	1245,41	32,95	106,923	112
		2	1245,26	32,95	106,925	220
		6	1245,49	32,95	106,928	1772
XROI	Dynamic	2	1280,55	33,26	114,340	780
		6	1285,36	33,27	114,343	6884
	Static	2	1419,17	33,63	113,012	252
		6	1419,4	33,63	114,426	1772

larger bitrate is needed when using FMO type 6 it should be mentioned that this type has some advantages over type 2. First, less decoding time is needed (not shown), due to the fact that the decoder does not have to generate the FMO map, but instead receives all necessary information from the picture parameter set. Secondly, other structures than typical symmetric squares can be generated with type 6. Note that the PSRN of the xROI implementations is larger than those with the basic ROI implementation without xROI. This is normal since the XROI area is quantized more coarsely in the examples without an xROI, where it is simply part of the non-ROI.

19.4.2 ROI Shape and QP Set Comparison

To find out which shapes (see Table 19.2 for an overview of shapes that were implemented) are useful in certain circumstances, multiple ROI shapes, all having approximately the same number of MBs in the ROI, have been tested with several different QP sets applied to them. The same, changing ROI positions were used in each test.

The simulations were done once for each combination of a Shape and QP set because no variable parameters, like encoding time, are taken into account. Simulation with the same combination will always result in the same output, which means the same bitrate and same resulting sequence. This sequence will thus generate the same PSNR and SSIM measurements. The free "MSU Video Quality Measure Version 1.0" tool was used for the SSIM measurements [2]. The tests were again done using the "Football" sequence. For the SSIM calculations, we pre-encoded two sequences, one with QP 15 and one with QP 25. This QP values were applied on the whole picture (so there is no (x)ROI).

Table 19.2. Implementation Overview (x denotes: not implemented)

Shapes	FMO type 3	FMO type 2	FMO type 6	tested ROI size	number of MBs
Square	✓	✓	✓	4	81
Rectangle	x	✓	✓	8 * 2	85
Diamond	x	x	✓	6	85
Octagon	x	x	✓	5	81
Interactive	x	✓	✓		
Extensible	x	✓	✓		

One of the most interesting things to see is that for some QP-sets the bitrate is higher than for others, while the quality measurements are worse as can be seen on Fig. 19.6

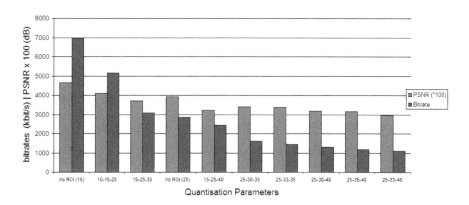

Fig. 19.6. Bitrates and PSNR for 'Square' with different QP sets

For example, the image when using QP set (15-25-40) is worse for every quality measurement compared to the images when using QP sets (25-30-35) or (25-33-35) while the last two bitrates are lower. This also can be seen in Table 19.3. One would expect that when higher bitrates are obtained, the resulting quality would become better. Explanation can be found in the logarithmic nature of the quantization parameters. The difference between QP 15 and QP 25 is visually less noticeable (both for humans as for quality measurements) than it is between QP 35 and QP 40. Thus the first difference will not influence PSNR and SSIM much.

The shape that should be chosen will depend mainly on the content of the picture. This means that the QP set can only be controlled to influence the bitrate and the perceived image quality. Additionally, we compared different ROI shapes. Depending on the situation, one will be favored above the other.

Table 19.3. Summarized Table

	Square no xROI		Rectangle no xROI	
	15-25-40	25-30-35	15-25-40	25-30-35
Bitrate (kbps)	2646,52	1730,44	2534,08	1624,07
PSNR	28,69	31,02	28,32	30,71
SSIM 15	0,82218	0,88346	0,79092	0,87951
SSIM 25	0,82986	0,9003	0,80597	0,88312

Our experimental tests showed that a rectangle without xROI will have the lowest bitrate. For a sequence like "BUS", this is a good approach, but when the ROI is e.g. a head, it is not desirable to see only the eyes. The reason why the square delivers better results for a given QP set, even though it has more MBs encoded with a lower QP is due to the content inside thos MBs. We also discovered that for the majority of QP sets we tested, the square without xROI always has a lower bitrate than the diamond or the octagonal. But these xROI solutions usually generate the highest bitrates compared to the other xROI bitrates. This is mainly caused by the high number of MBs contained in the xROI.

From these results we can formulate the following conclusion: When less bandwidth is available, lowering the QP of the non-ROI should be done first. Care should be taken that a homogeneous QP set is preserved, meaning that the 3 QPs are in a relatively close interval. This way the overall picture quality is perceived to be better. This is in contrast to QP sets (15-25-40) and (25-30-40) (not shown in table) where bitrates are higher and PSNR is less in comparison with homogeneous QPs in the same range. After increasing the non-ROI QP, it is recommended to increase the xROI QP. If necessary, QPs can still be increased or the size of the ROI can be reduced. All this will result in a reduction of bandwidth. With a noticeable, but still acceptable difference in quality, we can reduce the bandwidth of 2.5 Mbps (with the global QP set to 25) to approximately 1.3 Mbps (QP set 25-33-35). Apart from this, different shapes will of course give different results.

19.5 Conclusions

The need for ROI coding in mobile video applications or other scalability techniques is becoming apparent. H.264, a standard that is becoming very popular, is a typical coder that might be used in these mobile applications. ROI coding in this codec is relatively straightforward thanks to some very interesting features in the standard, such as usage of slices and flexible macroblock ordering. Existing research is mainly focused on reducing bandwidth and not on the user's viewing experience. Our research was focused on ROI-shape comparison and the introduction of an extensible ROI to make the transition of the ROI to non-ROI more smooth. Extensive testing of different shapes and FMO types enabled us to discover some very useful results when it comes to choosing the kind of ROI implementation to be used.

The major differences between the FMO types are with respect to the picture parameter settings and the ordering of the MBs. FMO type 3 sends less picture parameter set information compared to the other FMO types, but the different ordering of the MBs can lead to an increased bitrate. Additionally, the RoiPos cannot be moved and no xROI can be applied when box-out is used. Therefore, we advise not to use the box-out ROI model (FMO type 3).

When a complex algorithm can be implemented - it does not necessarily lead to larger encoding times - we advise to use FMO type 6. This ROI model can be a predefined shape, an amorph shape, or just a selection of MBs generated by an algorithm or based on an activity map [3].

Lower bitrates are not always guaranteed when using the same number of MBs in the same area as when using FMO type 2. When a more accurate shape can be constructed, bandwidth will be reduced because less MBs will be in high quality. When just a basic approach of ROI functionality can be implemented, less attention must be given to an accurate ROI detection model. For simple applications, foreground-background FMO mapping should be used, a combination of the latter and explicit or only explicit FMO mapping is recommended in more advanced applications.

References

1. H.264 reference software.
2. Msu video quality measure version 1.0.
3. Pierpaolo Baccichet, Xiaoqing Zhu, and Bernd Girod. Network-aware h.264/avc region-of-interest coding for a multi-camera wireless surveillance network. *Proc. Picture Coding Symposium, (PCS-06), Beijing, China, April*, 2006.
4. Yves Dhondt, Peter Lambert, Stijn Notebaert, and Rik Van de Walle. Flexible macroblock ordering as a content adaptation tool in h.264/avc. *Proc. of SPIE Vol. 6015 601506-1*, 2006.
5. ISO/IEC and JVT. Text of iso/iec 14496-10:2004 advanced video coding. *ISO/IEC JTC1/SC29/WG11/N6359*, page 280, 2004.
6. Iain E. G. Richardson. *H/264 and MPEG-4 Video Compression*. Wiley, 2003.

Rough Sets-Based Image Processing for Deinterlacing

Gwanggil Jeon[1] and Jechang Jeong[1]

Department of Electronics and Computer Engineering, Hanyang University,
17 Haengdang-dong, Seongdong-gu, Seoul, Korea
{windcap315,jjeong}@ece.hanyang.ac.kr

Summary. This chapter includes the rough sets theory for video deinterlacing that has been both researched and applied. The domain knowledge of several experts influences the decision making aspects of this theory. However, included here are a few studies that discuss the effectiveness of the rough sets concept in the field of engineering. Moreover, the studies involving a deinterlacing system that are based on rough sets have not been proposed yet. This chapter introduces a deinterlacing method that will reliably confirm that the method being tested is the most suitable for the sequence. This approach employs a reduced database system size, which contains the essential information for the process. Decision making and interpolation results are presented. The results of computer simulations show that the proposed method outperforms a number of methods that are presented in literature.

20.1 Introduction

The current analog television standards, such as NTSC, PAL, and SECAM, are based on interlaced scanning formats. Because the video industry is transitioning from analog to digital, video processing equipment increasingly needs to transition from analog to digital as well. Thus, the demand for progressive material will increase, which causes a directly proportional increase in the demand for video processing products with high quality deinterlacing. Deinterlacing methods can be roughly classified into three categories: spatial domain methods [1],[2], which use only one field; temporal domain methods [3], which use multiple fields; and spatio-temporal domain methods [4]. The most common method in the spatial domain is Bob [2], which is used on small LCD panels. However, the vertical resolution is halved, and this causes the image to have jagged edges. Weave is the most common method in the temporal domain [3]. However, this method gives motion artifacts. There exist many edge direction based interpolation methods. The edge line average (ELA) algorithm was proposed to interpolate pixels along the edges in the image [1]. Oh et al. propose a spatio-temporal line average (STELA) algorithm. ELA utilizes only the spatial domain information. However, the amount of data limits the interpolation by causing missed pixels at complex and motion regions. Thus, STELA was proposed in order to expand the window to include the temporal domain. Generally, various features offer several attributes for the nature of a sequence. However, sometimes the attributes become too much to make essential rules. Although some rules are decided, even human experts are unable to believe the rules. Thus, the conventional deinterlacing method cannot be applied to build an expert system. In order to create an expert system, rough sets theory is applied to classify the deinterlacing method. In this theory, prior knowledge of the rules is not required, but rather the rules are automatically discovered from a database. Rough sets theory provides a formal and robust method of manipulating the roughness in information systems [5]. It has been applied to several areas including knowledge discovery [6],[7],[8],[9], feature selection [10], clustering [11], image recognition and segmentation [12],[13],[14], quality evaluation [15], and medical image segmentation [16],[17],[18]. It has proved its advantage in real world applications, such as semiconductor manufacturing [19], landmine classification [20], fishery applications [21], and power system controllers [22]. Rough sets theory has been used in imaging, but its application in video deinterlacing has yet to be investigated. This chapter presents a decision making algorithm that is based on rough sets theory for video deinterlacing. The operation of a decision in the deinterlacing method is intrinsically complex due to the high degree of uncertainty and the large number of variables involved. The analysis performed by the operator attempts to classify the operational state of the system in one of four states: plain-stationary region, complex-stationary region, plain-motion region, or complex-motion region. The proposed rough sets deinterlacing (RSD) algorithm employs four deinterlacing methods: Bob [2], Weave [3], ELA [1], and STELA [4]. In Section 20.2, the basic concepts of the rough sets theory are discussed. In Section 20.3, the proposed rough sets deinterlacing algorithm is described. In Section 20.4, the experimental results and performance analysis are provided to show the feasibility of the proposed approach. These results are compared to well-known, pre-existing deinterlacing methods. Finally, conclusions are presented in Section 20.5.

20.2 Basic Concepts of Rough Sets Theory

Rough sets, introduced by Pawlak et al., is a powerful tool for data analysis and characterizing imprecise and ambiguous data. It has successfully been used in many application domains, such as machine learning and expert systems [5].

20.2.1 Preliminary

Let $U \neq \emptyset$ be a universe of discourse and X be a subset of U. An equivalence relation, R, classifies U into a set of subsets $U/R = \{X_1, X_2, \cdots, X_n\}$ in which the following conditions are satisfied:

$$X_i \subseteq U, X_i \neq \emptyset \, for \, any \, i$$
$$X_i \cap X_j \neq \emptyset \, for \, any \, i, j \qquad (20.1)$$
$$\bigcup_{i=1,2,\cdots,n} X_i = U$$

Any subset X_i, which is called a *category*, *class*, or *granule*, represents an equivalence class of R. A category in R containing an object $x \in U$ is denoted by $[x]_R$. For a family of equivalence relations $P \subseteq R$, an *indiscernibility relation over* P is denoted by $IND(P)$ and is defined as follows:

$$IND(P) = \bigcap_{R \in P} IND(R) \qquad (20.2)$$

The set X can be divided according to the basic sets of R, namely a *lower approximation set* and *upper approximation set*. Approximation is used to represent the roughness of the knowledge. Suppose a set $X \subseteq U$ represents a vague concept, then the $R - lower$ and $R - upper$ approximations of X are defined.

$$\underline{R}X = \{x \in U : [x]_R \subseteq X\} \qquad (20.3)$$

Equation 20.3 is the subset of all X, such that X belongs to X in R, is the lower approximation of X.

$$\overline{R}X = \left\{x \in U : [x]_R \cap X \neq \emptyset\right\} \qquad (20.4)$$

Equation 20.4 is the subsets of all X that possibly belong to X in R, thereby meaning that X may or may not belong to X in R. The lower approximation $\underline{R}X$ contains sets that are certainly included in X, and the upper approximation $\overline{R}X$ contains sets that are possibly included in X. $R-positive$, $R - negative$, and $R - boundary$ regions of X are defined respectively as follows:

$$POS_R(X) = \underline{R}X, \, NEG_R(X) = U - \overline{R}X, \, BN_R(X) = \overline{R}X - \underline{R}X \qquad (20.5)$$

Fig. 20.1 shows the $R - lower$ and $R - upper$ approximations of the family X, and three kinds of regions: $R - negative$ region of X, $R - boundary$ region of X, and $R - positive$ region of X.

20.2.2 Reduct and Core

In rough sets theory, a decision table is used for describing the object of universe. The decision table consists of two dimensional tables. Each row is an object, and each column is an attribute. Attributes can be divided into either a condition attribute or a decision attribute. However, the entire condition attribute may not be essential. Because there may exist surplus attributes

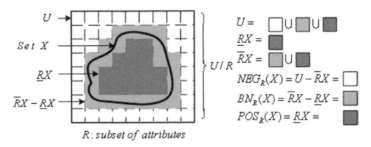

Fig. 20.1. Definition of R-approximation sets and R-regions.

and this excess can be eliminated, correct classification is guaranteed. Rough sets theory classifies the attributes in the decision table into three types according to their roles in the decision table: *core* attributes, *reduct* attributes, and *superfluous* attributes. Here, the minimum condition attribute set can be received, which is called *reduction*. One decision table might have several different reductions simultaneously. The intersection of the reductions is the *core* of the decision table and the attribute of the core is the important attribute that influences attribute classification. Suppose R is a family of equivalence relations. The reduct of R, $RED(R)$, is defined as a reduced set of relations that conserves the same inductive classification of set R. The core of R, $CORE(R)$, is the set of relations that appears in all reduct of R, i.e., the set of all indispensable relations to characterize the relation R. Generally, rough sets theory provides several advantages as a data mining tool. Rough sets theory provides a consistent mathematics tool that can rigidly deal with data classification problems. Thanks to data reduction and data core, useful characteristics can be selected and are sufficient to express the data. Reducing the amount of data lessens computational time as well. Rough sets theory contains a form that represents a model of knowledge. This model becomes a set of equivalent relations, which clarifies the mathematical meaning. The rules of classification are obtained by analyzing and processing with mathematical methods. Finally, rough sets theory does not require extra information, which allows for practical application in conjunction with professional knowledge. This improves attribute reduction and produces specification based decisions of a high quality. The problem of identifying the application space is similar to that of identifying redundant attributes and eliminating the redundancy. Hence, all the attributes with no useful information are removed.

20.3 Heuristics for Information Acquisition and Its Application

Nearly all feature values are continuous data, and are demonstrated to be unsuitable for the extraction of concise symbolic rules. Simultaneously, the conditional rules have poor predictions. Hence, the original analog data must be transformed into normalized discrete data. The process of transforming data sets with continuous attributes into input data sets with discrete attributes is called discretization. Continuous data must be converted to discrete intervals, in which each interval is represented by a label. Discretization not only reduces the complexity and volume of the data set, but also serves as an attribute filtering mechanism. In this chapter, it is assumed that an image can be classified according to four main parameters: $SMDW$, $TMDW$, SD, and TD (5-6). β is an amplification factor that affects the size of membership functions resulting in $TMDW$ and $SMDW$ varying between 0 and 255. N_{W_T} and N_{W_S} each provide 6, and $x(i, j, k)$ denotes the intensity of a pixel, which will be interpolated in our work. i refers to the column number, j refers to the line number, and k refers to the field number.

$$SMDW = \frac{\left[max_{(i,j,k)\in W_S} x(i,j,k) - min_{(i,j,k)\in W_S} x(i,j,k) \right] \times N_{W_S}}{\sum_{(i,j,k)\in W_S} x(i,j,k)} \times \beta$$

(20.6)

$$SD = |x(i, j-1, k) - x(i, j+1, k)|$$

(20.7)

$$TMDW = \frac{\left[max_{(i,j,k)\in W_T} x(i,j,k) - min_{(i,j,k)\in W_T} x(i,j,k) \right] \times N_{W_T}}{\sum_{(i,j,k)\in W_T} x(i,j,k)} \times \beta$$

(20.8)

$$TD = |x(i, j, k-1) - x(i, j, k+1)|$$

(20.9)

The spatial domain maximum difference over the window ($SMDW$) parameter and the temporal domain maximum difference over the window ($TMDW$) parameter represent the spatial and temporal entropy. Spatial difference (SD) or temporal difference (TD) is the pixel difference between two values across the missing pixel in each domain. The continuous values of the features have been discretized into a symbol table. We assume that the pixels with low $SMDW$ or SD are classified within the plain area and that the others are classified within the complex area. Furthermore, the pixels with low $TMDW$ or TD are classified within the static area, and the remaining pixels are classified in the motion area. Based on this classification system, a different deinterlacing algorithm is activated, in order to obtain the best performance. Twelve pixels around the missing pixel must be read before attributes may be extracted. Then, the extracted attributes are normalized at the position of each missing pixel. The step of categorization of the attribute involves converting the attributes from numerical to categorical. Data may be lost during the conversion from analog to digital information. Neither of the methods that are simply based on frequencies, nor those that are based on boundaries, are optimal. Instead, the numerical range is determined according to the frequencies of each category boundary.

abcd	total	d_B	d_W	d_E	d_T	m	abcd	total	d_B	d_W	d_E	d_T	m
1111(1)	8563	4965	2778	650	170	B	2111 (5)	1189	390	432	247	120	W
1112	287	171	46	59	11	-	2112	66	32	12	11	11	-
1113(2)	1663	823	223	526	91	B	2113	837	186	191	260	200	-
1121(3)	1065	107	711	147	100	W	2121	121	8	62	30	21	-
1122	143	44	57	12	30	-	2122	23	3	6	6	8	-
1123	230	47	73	41	69	-	2123	55	16	15	10	14	-
1131 (4)	1344	261	614	283	186	W	2131 (6)	2168	296	784	541	547	W
1132	149	34	45	31	39	-	2132	195	32	58	46	59	-
1133 (7)	1678	357	355	636	330	E	2133 (10)	5411	780	1386	1541	1704	T
1211	300	126	99	41	34	-	2211	670	195	146	187	142	-
1212	14	6	3	4	1	-	2212	30	11	7	8	4	-
1213	430	137	83	143	67	-	2213 (8)	1877	387	374	584	532	E
1221	86	13	34	24	15	-	2221	74	7	27	25	15	-
1222	8	2	4	0	2	-	2222	10	3	2	2	3	-
1223	103	13	46	24	20	-	2223	89	18	30	16	25	-
1231	372	73	116	89	94	-	2231 (11)	2696	452	670	736	838	T
1232	12	4	3	1	4	-	2232	91	17	24	24	26	-
1233 (9)	1431	214	279	305	633	T	2233 (12)	16570	2613	3792	4810	5355	T

Table 20.1. Set of the selected method (m) corresponding to each pattern.

The classification of each state is made according to an expert, and four possible regions can be selected for the decision making for the video deinterlacing system: plain-stationary region, complex-stationary region, plain-motion region, or the complex-motion region. The first step of the algorithm is to redefine the value of each attribute according to a certain metric. The set of all possible decisions is listed in Table 20.1. Let Table 20.1, where the information system proposed is composed of $R = \{a, b, c, d, m | (a, b, c, d) \rightarrow (m)\}$. This table is a decision table in which a,b,c, and d are condition attributes, whereas m is a decision attribute. Using these values, a set of examples can be generated. The attribute m represents the expert's decisions, which are the following:

m : B-Bob method; W-Weave method; E-ELA method; T-STELA method.
a : 1-SMALL (SMDW$\leq$5); 2-LARGE (SMDW>5)
b : 1-SMALL (TMDW$\leq$5); 2-LARGE (TMDW>5)
c : 1-SMALL (SD$\leq$2); 2-MEDIUM (2<SD$\leq$5); 3-LARGE (5<SD)
d : 1-SMALL (TD$\leq$2); 2-MEDIUM (2<TD$\leq$5); 3-LARGE (5<TD)

Average picture of the Foreman sequence has been employed as a training image (for the 2nd to the 298th). It was found that 12 sets out of 36 sets have more than 1000 pixels. The total pixel amount in the 12 sets is more than 93% of all pixels. The other 24 sets have at most 7% of all pixels, and will provide the complexity to the system. Thus, it is unsuitable to generate rules to classify the other 24 sets. Furthermore, according to a hidden Markov model, since the correlation of the image is more than 0.95, a set with a small number of pixels will be considered an edge region. The STELA method is suitable for that region. Each twelve row in Table 20.1 has three bold characters, which represent sorted number according to decision attributes m_i (i: B, W, E, and T), which is shown in Table 20.2, the total number of pixels in each row, and the largest number among four methods, respectively. Then a decision must be made as to which method is the most suitable for the missing pixel. The difference between the real value and the Bob interpolated value is regarded as d_B. In the same manner, d_W, d_E, and d_T are obtained. Finally, the most suitable method is selected as the one with the smallest value among the four differences. When the differences are equal, priority is given to the smallest complexity: Bob, Weave, ELA, and STELA. Generally, Bob exhibits no motion artifacts and has minimal computational requirements. However, the input for vertical resolution is halved before the image is interpolated, thus reducing the detail in the progressive image. The Weave process results in no degradation of static images. However, the edges exhibit significant serrations, which is an unacceptable artifact in a broadcast or professional television environment. Both of these techniques require less complexity for interpolating a missing pixel. The processing requirements for ELA or STELA are higher than that of Bob or Weave yet with the advantage of higher output image quality. Rough set theory offers one mathematic method that can strictly treat data classification problems. The idea behind the knowledge base reduction is a simplification of Table 20.1. The algorithm that provides the reduction of conditions is represented by the following steps: 1) Removing dispensable attributes. 2) Finding the core of the decision table. 3) Associate a table with reduct value. 4) Extract possible rules. To simplify decision table, the reduction of the set of condition categories is necessary to define the decision categories. By removing attributes a in Table 20.2(a), a decision table, Table 20.2(b), is provided. However, this is inconsistent because Table 20.2(b) contains the following pairs (20.10), (20.11) of inconsistent decision rules:

$$(rule\,1 : b_1c_1d_1 \rightarrow m_B)\,and\,(rule5 : b_1c_1d_1 \rightarrow m_W) \qquad (20.10)$$

$$(rule\,7 : b_1c_3d_3 \rightarrow m_E)\,and\,(rule10 : b_1c_3d_3 \rightarrow m_T) \qquad (20.11)$$

Thus, the attributes a cannot be removed. In the same manner, it has been observed that all attributes are indispensable. This indicates that none of the condition attributes can be removed from Table 20.2(a). Hence the set of condition attributes is m-independent. The next step is to check where some elementary condition categories can be eliminated, i.e., some superfluous values of condition attributes in Table 20.2(a). For example, in the sixth decision rule $a_2b_1c_3d_1 \rightarrow m_W$ values b_1 and d_1 are core values because the rules $a_2b_1d_1 \rightarrow m_W$ and $b_1c_3d_1 \rightarrow m_W$ are true, whereas the rules $a_2b_1c_3 \rightarrow m_W$ and $a_2c_3d_1 \rightarrow m_W$ are false. The core values of each decision rule in Table 20.2(a) are given in Table 20.2(c). Now, all m-reducts of condition elementary

	(a)						(b)					(c)				

U	a	b	c	d	m	U	b	c	d	m	U	a	b	c	d	m
1	1	1	1	1	B	1	1	1	1	B	1	1	-	1	-	B
2	1	1	1	3	B	2	1	1	3	B	2	-	-	-	-	B
3	1	1	2	1	W	3	1	2	1	W	3	-	-	2	-	W
4	1	1	3	1	W	4	1	3	1	W	4	-	-	3	1	W
5	2	1	1	1	W	5	1	1	1	W	5	2	-	-	-	W
6	2	1	3	1	W	6	1	3	1	W	6	-	1	-	1	W
7	1	1	3	3	E	7	1	3	3	E	7	1	1	3	3	E
8	2	2	1	3	E	8	2	1	3	E	8	-	-	1	-	E
9	1	2	3	3	T	9	2	3	3	T	9	-	2	-	-	T
10	2	1	3	3	T	10	1	3	3	T	10	2	-	-	3	T
11	2	2	3	1	T	11	2	3	1	T	11	-	2	-	-	T
12	2	2	3	3	T	12	2	3	3	T	12	-	-	3	-	T

Table 20.2. (a) Set of deinterlacing system; (b) Removing attributes a from Table 2; (c) Core of the attributes.

categories or the reduct values of the condition attributes of each decision rule can be computed. In order to find reducts of the decision rules, the core values of each decision rule, such as values of condition attributes of the rule, need to be added so, that the predecessor of the rule is independent and the whole rule is true. For example the sixth decision rule $a_2 b_1 c_3 d_1 \rightarrow m_W$ has two reducts $a_2 b_1 d_1 \rightarrow m_W$ and $b_1 c_3 d_1 \rightarrow m_W$, since both decision rules are true and predecessor of each decision rule is independent. The results of each decision rule in Table 20.2(a) are listed in Table 20.3(a). In order to find the minimal decision algorithm, all superfluous decision rules must be removed from the table. Table 20.3(b) shows the final essential decision rules. The final results, presented in Table 20.3(b), can be rewritten as a minimal decision algorithm in normal form. Combining the decision rules into one decision class provides the following decision algorithm.

$$
\begin{aligned}
&if((a_1 c_1) \vee (b_1 c_1 d_3))\ m_B \\
&else\,if(c_2 \vee ((a_1 \vee b_1)c_3 d_1) \vee ((b_1 c_1 \vee b_1 d_1 \vee c_1 d_1)a_2))\ m_W \\
&else\,if((b_2 \vee a_2 d_3)c_1 \vee a_1 b_1 c_3 d_3)\ m_E \\
&else\ m_T
\end{aligned}
\qquad (20.12)
$$

20.4 Experimental Results

In this section, objective quality and computational time are compared. Various simulations have been carried out, and natural images have been tested for the verification. The proposed algorithm was implemented on a Pentium

(a)   (b)

U	a	b	c	d	m	U	a	b	c	d	m
1	1	1	1	x	B	1(1,1',2)	1	x	1	x	B
1'	1	x	1	1	B	2(2')	x	1	1	3	B
2	1	x	1	x	B	3(3)	x	x	2	x	W
2'	x	1	1	3	B	4(4)	1	x	3	1	W
3	x	x	2	x	W	5(4',6)	x	1	3	1	W
4	1	x	3	1	W	6(5)	2	1	1	x	W
4'	x	1	3	1	W	7(5',6)	2	1	x	1	W
5	2	1	1	x	W	8(5")	2	x	1	1	W
5'	2	1	x	1	W	9(7)	1	1	3	3	E
5"	2	x	1	1	W	10(8)	x	2	1	x	E
6	2	1	x	1	W	11(8')	2	x	1	3	E
6'	x	1	3	1	W	12(9)	1	2	x	x	T
7	1	1	3	3	E	13(9',11,12)	x	2	3	x	T
8	x	2	1	x	E	14(10)	2	1	x	3	T
8'	2	x	1	3	E	15(10',12')	2	x	3	3	T
9	1	2	x	x	T	16(11')	x	2	x	1	T
9'	x	2	3	x	T						
10	2	1	x	3	T						
10'	2	x	3	3	T						
11	x	2	3	x	T						
11'	x	2	x	1	T						
12	x	2	3	x	T						
12'	2	x	3	3	T						

Table 20.3. (a) All reducts decision values; (b) Final sets of deinterlacing methods.

IV/2.80 GHz computer. For the objective performance evaluation, five CIF video sequences were selected to challenge the five algorithms for ELA, Bob, Weave, STELA, and the proposed method. Table 20.4 shows the PSNR result of different deinterlacing methods for various sequences. The results show that the proposed method demonstrates the 2nd best objective performance compared to the other conventional methods, in terms of PSNR. It also shows that the proposed method has slightly less computational CPU time than the ELA method.

For a subjective performance evaluation, the 171st frame of the CIF Table Tennis sequence was adopted. The subjective views of the video sequences are shown in Fig. 20.2. The wall, shadow, and the surface on the table are classified into a plain-stationary region, the picture on the wall and the edge of the table into a complex-stationary region, the shirt and the pants into a plain-motion region, and the ball, right hand with racket, and left hand into a complex-motion region. Fig. 20.2(b) and (c) shows no motion artifacts in

Method	Sequences (PSNR, computational CPU time)				
	Akiyo	Flower	Mobile	News	Table Tennis
Bob	39.69 dB	22.40 dB	25.49 dB	33.66 dB	32.01 dB
	12.71 ms	15.27 ms	13.74 ms	12.92 ms	13.76 ms
Weave	40.67 dB	20.31 dB	23.36 dB	36.29 dB	24.75 dB
	11.30 ms	12.38 ms	13.57 ms	11.68 ms	13.18 ms
ELA	37.68 dB	21.93 dB	23.34 dB	31.53 dB	31.23 dB
	28.74 ms	28.83 ms	31.51 ms	29.47 ms	29.07 ms
STELA	44.65 dB	22.99 dB	27.26 dB	39.28 dB	31.58 dB
	42.96 ms	44.43 ms	48.39 ms	44.07 ms	44.74 ms
Proposed	44.26 dB	22.53 dB	26.73 dB	39.02 dB	31.08 dB
	22.34 ms	23.63 ms	25.06 ms	22.92 ms	23.61 ms

Table 20.4. Results of different interpolation methods for five CIF sequences (Unit = dB, ms).

the motion region. However, the input vertical resolution is halved before the image is interpolated, thus reducing the detail in the progressive image. While Weave results in no degradation of static region, the moving hand exhibits significant serrations, see Fig. 20.2(d). STELA gradually reduces the vertical detail as the temporal frequencies increase, as shown in Fig. 20.2(e). The vertical detail from the previous field is combined with the temporally shifted current field, indicating that some motion blur occurred. Fig. 20.2(f) shows the proposed RSD utilized image. RSD has slightly less quality than that of STELA. RSD causes some degradation at the region around the picture on the wall, yet only with the 52.34% of complexity of STELA algorithm. From the experiment results, it is observed that the proposed RSD algorithm has good objective and subjective qualities for different sequences, with a low computational CPU time required to achieve the real-time processing.

20.5 Conclusion

This chapter describes an application of rough set to feature selection and reduction in deinterlacing systems. Few studies exist that discuss the effectiveness of the rough set concept in the field of engineering, where the domain knowledge of experts plays a key role in determining deinterlacing methods. Moreover, the studies involving deinterlacing systems that are based on the rough set method have not been proposed yet. This chapter presents a novel deinterlacing approach using a reduced size of database, which keeps only the essential information to the process. Decision making and interpolation results are presented. The results of computer simulations show that the proposed method outperforms a number of methods in literature.

(a) Original

(b) ELA

(c) Bob

(d) Weave

(e) STELA

(f) Proposed method

Fig. 20.2. Subjective quality comparison of the 171st Table Tennis CIF sequence.

References

1. T. Doyle, "Interlaced to sequential conversion for EDTV applications," in Proc. 2nd Int. Workshop Signal Processing of HDTV, pp. 412-430 Feb. 1990
2. E. B. Bellers and G. de Haan, "Advanced de-interlacing techniques," in Proc. ProRisc/IEEE Workshop on Circuits, Systems and Signal Processing, Mierlo, The Netherlands, Nov. 1996, pp. 7-17
3. P. L. Swan, "Method and apparatus for providing interlaced video on a progressive display," U.S. Patent 5 864 369, Jan. 26, 1999

4. H. -S. Oh, Y. Kim, Y. -Y. Jung, A. W. Morales, and S. -J. Ko, "Spatio-temporal edge-based median filtering for deinterlacing," IEEE International Conference on Consumer Electronics, pp. 52-53, 2000

5. Z. Pawlak - "Rough Sets - Theoretical Aspects of Reasoning about Data," Klumer Academic Publishers, 1991

6. Q. Wu, X. Huang, and S. Van, "Multi-knowledge for robot to identify environments," in Proc. WCICA 2004, Hangzhou, China, 2004, pp. 4840 - 4845 Vol.6.

7. M. Li, X. -F. Zhang, " Knowledge entropy in rough set theory," in Proc. ICMLC 2005, Shanghai, China, pp. 1408 - 1412 vol.3

8. M. Liu, Y. He, H. Hu, and D. Yu, "Dimension reduction based on rough set in image mining," in Proc. CIT 2004, Chennai, India, pp. 39 - 44

9. X. -F. Zhang, F. -Z. Zhang, and Y. -S. Zhao, "Generalization of RST in ordered information table," in Proc. ICMLC 2005, Guangzhou, China, 2005, pp. 2027 - 2032 Vol. 4

10. L. Pan, H. Zheng, S. Nahavandi, "The application of rough set and Kohonen network to feature selection for object extraction," in Proc. ICMLC 2003, Xi'an, China, 2003, pp. 1185 - 1189 Vol.2

11. J. W. Grzymala-Busse, "LERS - A system for learning from examples based on rough sets," in R. Slowinski (Ed.) Intelligent Decision Support. Handbook of Applications an Advances of the Rough Set Theory, Kluwer Academic Publishers, Dordrecht, 1992

12. Xiaohua Hu, "Using rough sets theory and database operations to construct a good ensemble of classifiers for data mining applications," in Proc. ICDM 2001, San Jose, California, 2001, pp. 233 - 240.

13. A. Mohabey and A. K. Ray, "Rough set theory based segmentation of color images," in Proc. NAFIPS 2000, Atlanta, GA, pp. 338 - 342

14. S. Mitatha, K. Dejharn, F. Chevasuvit, B. Chankuang, and W. Kasemsiri, "Experimental results of using rough sets for printed Thai characters recognition," in Proc. TENCON 2001, Cairns, Thailand, pp. 331 - 334 vol.1

15. X. Wu and Q. Wang "Application of rough set attributes reduction in quality evaluation of dissertation," in Proc. ICGC 2006, Atlanta, GA, pp. 562 - 565

16. Y. Peng, G. Liu, T. Lin, and H. Geng, "Application of rough set theory in network fault diagnosis," in Proc. ICITA 2005, Hangzhou, China, 2005, 556 - 559 vol.2

17. Y. -H. Xu, W. -J. Jiang, and Y. -S. Xu, "Research on extracting medical diagnosis rules based on rough sets theory," in Proc. ICMLC 2005, Guangzhou, China, 2005, pp. 3713 - 3718 Vol. 6.

18. S. Hirano, X. Sun, and S. Tsumoto, "Dealing with multiple types of expert knowledge in medical image segmentation: a rough sets style approach," in Proc. FUZZ-IEEE'02, Honolulu, Hawaii, 2002, pp. 884 - 889

19. A. Kusiak, "Rough Set Theory: A data mining tool for semiconductor manufacturing," IEEE Trans. Electronics Packaging Manufacturing, vol. 24, no. 1, pp. 44-50, Jan. 2001

20. A. K. Agrawal and A. Agarwal, "Rough logic for building a landmine classifier," in Proc. ICNSC 2005, Tucson, AZ, pp. 855 - 860

21. F. Su, C. Zhou, and W. Shi, "Geoevent association rule discovery model based on rough set with marine fishery application," in Proc. IGARSS 2004, Anchorage, Alaska, 2004, pp. 1455 - 1458 vol.2.

22. L. Torres, "Application of rough sets in power system control center data mining," in Proc. PESW 2002, New York, NY, pp. 627 - 631 vol.1.

Intersubband Reconstruction
of Lost Low Frequency Coefficients
in Wavelet Coded Images

Joost Rombaut, Aleksandra Pižurica, and Wilfried Philips

Ghent University – TELIN – IPI – IBBT,
St-Pietersnieuwstraat 41, B-9000 Ghent, Belgium,
Telephone: +32 9 264 95 30, Fax: +32 9 264 42 95,
Email: jorombau@telin.ugent.be, Web: http://telin.ugent.be/~jorombau/

Summary. In packet switched networks, packets may get lost during transmission. As these networks are more and more used for image and video communication, there is a growing need for efficient reconstruction algorithms. In wavelet coded images, the lost coefficients are typically replaced by zeros. This results in annoying black holes in the received image, mainly due to the loss of the low frequency content. In this chapter, we present a novel locally adaptive interpolation method for the reconstruction of the lost low frequency coefficients. We interpolate a lost low frequency coefficient from its four neighbors, and we determine the interpolation weights by the energy of the corresponding coefficients in the high frequency subbands.

Compared to older methods of similar complexity, the proposed scheme estimates the lost coefficients much better: the Peak Signal to Noise Ratio is increased with up to 4.3 dB. The results demonstrate a significant improvement of the visual quality.

Key words: passive error concealment, image reconstruction, wavelet coding, packet loss, error concealment, image communication

21.1 Introduction

Data loss arises often in packet switched networks due to network congestions. This is an especially important problem in case of compressed data, where the loss of a single bit may make the rest of the data stream unusable. In typical, non urgent network applications such as email, the data is often protected (e.g., by forward error correction) or in case of data loss, a packet can be retransmitted. These techniques are called *Active Error Concealment*. A good overview is given in [1]. In certain applications, such as real time video communication, the retransmission of a packet may be too slow and hence not tolerable, or in case of broadcasting, there simply may be no return channel. In these cases, *Passive Error Concealment*, i.e., postprocessing at the receiver, is necessary to achieve a high quality of the received video.

Passive error concealment exploits the redundancy in the image, using the correctly received data to reconstruct the lost information as well as possible. The data should be spread over different packets in order to make this reconstruction possible. This is achieved by the so-called packetization, which serves two purposes. Firstly, if some data gets lost during the transmission, then the beginning of a packet acts as a resynchronization point. Secondly, a good packetization spreads neighboring coefficients over different packets, ensuring in this way that the lost data can be estimated from its correctly received *neighboring* data. Examples of packetization techniques are parity based slicing [2] or a packetization based on the partitioning of the $\mathbb{Z}^2$ lattice [3]. In this chapter, we use the packetization strategy of [3], but any dispersive packetization strategy will work with our reconstruction algorithm.

We focus on wavelet based image and video coding. Loss of a packet of a wavelet coded image results in dark blobs. These blobs are mainly due to the loss of low frequency coefficients. As these coefficients contain most of the energy, they are the most important and should be reconstructed with most care. Although the wavelet transform tends to decorrelate the signal, there are substantial spatial dependencies between the coefficients, especially in the low-pass subband. These spatial dependencies can be used for the estimation of a lost coefficient. A bilinear interpolation [3] gives already good results, except near edges. The coefficients near edges are rapidly changing, and may be incorrectly estimated due to lack of correlation in at least one direction. The resulting errors are highly visible in the reconstructed image.

Different approaches exist to reconstruct the low frequency coefficients near edges more accurately. In [4], a lost low frequency coefficient is interpolated by fitting a cubic interpolative surface to the known coefficients. Correct edge placement is achieved by adapting the interpolation grid in horizontal and/or vertical direction according to the high frequency content. This method gives better results than the bilinear interpolation, but is also more complex and slower which may be less suited for low-end video clients such as portable devices with only a small processing capacity. In [5], the low frequency subband is repaired by a maximum a posteriori approach, using a Markov random field prior in each subband. The potential functions are adapted locally by estimating the edge characteristics based on the evolution of the coefficients across scales. This technique gives better results than the bilinear interpolation, but it requires much more computational effort. In our previous work [6], we proposed an interpolation technique where the lost low frequency coefficients are interpolated along the globally dominant correlation direction. This preferential direction is calculated by the sender and, after binning, sent to the receiver along with the wavelet coefficients. The receiver then adapts the interpolation weights according to this dominant correlation direction and the strength of this correlation. This method is faster than the fitting of a cubic interpolative surface of [4], and the maximum a posteriori approach of [5], but it has the following drawbacks. Firstly, the dominant correlation direction needs to be estimated at the sender side, which is not done by standard encoders. Secondly, this method takes into account only a globally dominant correlation direction, which may differ significantly from the locally dominant correlation properties at some positions in the image.

In this chapter, we propose a locally adaptive interpolation method with a complexity similar to that of the bilinear interpolation [3]. We estimate the locally optimal interpolation direction from the corresponding high frequency content (i.e., the frequential neighbors). In this way, we preserve the edge structures much better than standard schemes with constant interpolation weights. For low packet loss rates, the proposed interpolation method increases the Peak Signal to Noise Ratio (PSNR) with up to 4.3 dB compared to reconstruction methods of similar complexity. For high packet loss rates, the PSNR is increased with up to 1.85 dB.

In the next section we describe the proposed interpolation method. The reconstruction of the high frequency coefficients is presented in Sect. 21.3. Results and discussion are in Sect. 21.4, and in Sect. 21.5 we draw the conclusions and give some remarks about further work.

21.2 Reconstruction of Low Frequency Coefficients

Our reconstruction method is developed independently from the wavelet transform. We mainly tested it for the Symlet wavelet transform of order 4, but it can easily be extended to other types of wavelet transforms.

In the remainder, we use the following notation: LL^n denotes the low-pass subband (the scaling coefficients) at the decomposition level n; the wavelet coefficients are organized into the subbands LH^ℓ, HL^ℓ and HH^ℓ, which denote respectively horizontal, vertical and diagonal details at the decomposition level ℓ where $\ell \in \{1, \ldots, n\}$.

In the remainder of this section, we describe our reconstruction method for lost LL^n coefficients. The optimal interpolation weights are thereby estimated from the corresponding LH^n and HL^n coefficients. For clarity, we will omit the index n, which denotes the scale. The subscripts will denote the spatial position. For example, $LL_{i,j}$ denotes the scaling coefficient at spatial position (i, j).

21.2.1 Detection of the Local Correlation

In the proposed method, a lost coefficient $LL_{i,j}$ is estimated by adaptive weighted averaging in two directions: vertically (using the upper and lower coefficients $LL_{i-1,j}$ and $LL_{i+1,j}$), and horizontally (using the left and right coefficients $LL_{i,j-1}$ and $LL_{i,j+1}$):

$$\widehat{LL}_{i,j} = \alpha_{i,j}^V \left(LL_{i-1,j} + LL_{i+1,j} \right) + \alpha_{i,j}^H \left(LL_{i,j-1} + LL_{i,j+1} \right) . \quad (21.1)$$

As a neighbor of a lost coefficient may also be lost, we first interpolate lost neighbors by averaging out its correctly received neighboring coefficients.

The weighting factors for the vertical and horizontal direction, $\alpha_{i,j}^V$ and $\alpha_{i,j}^H$, are estimated locally at each spatial position (i,j). We relate these local interpolation weights to a measure of the local correlation in the corresponding directions. To estimate the local correlation, we use the high frequency subbands. Large magnitude coefficients in the HL subband indicate a vertical edge, and hence vertical correlation. Large magnitude coefficients in the LH subband indicate the opposite. We define:

$$E_{i,j}^{LH} = LH_{i-1,j}{}^2 + LH_{i,j}{}^2 \tag{21.2}$$

and

$$E_{i,j}^{HL} = HL_{i,j-1}{}^2 + HL_{i,j}{}^2 . \tag{21.3}$$

Intuitively, if $E_{i,j}^{LH} \gg E_{i,j}^{HL}$ then we would like to have $\alpha_{i,j}^H = 1/2$ and $\alpha_{i,j}^V = 0$, which means a horizontal interpolation. If $E_{i,j}^{LH} \ll E_{i,j}^{HL}$ then it would be best to choose $\alpha_{i,j}^H = 0$ and $\alpha_{i,j}^V = 1/2$, i.e., a vertical interpolation. In the following, we experimentally determine the optimal (in the mean squared error sense) relationship between the interpolation weights and the high frequency coefficient energies.

21.2.2 Optimal Interpolation Weights

Our experiments showed that $\alpha_{i,j}^H$ and $\alpha_{i,j}^V$ do not depend on the exact values of $E_{i,j}^{LH}$ and $E_{i,j}^{HL}$, but only on the ratio $E_{i,j}^{HL}/E_{i,j}^{LH}$. We define the high frequency energy ratio

$$R = \frac{E_{i,j}^{HL}}{E_{i,j}^{LH}} \tag{21.4}$$

as a measure of the local correlation direction. We calculated this energy ratio for all the low frequency coefficients (at level 3) from 146 different images. Then we quantized the obtained range of energy ratio values into 20 intervals. Next, the low frequency coefficients for which the energy ratio was within the same interval, were grouped together. For each of these groups of coefficients, we jointly optimized the interpolation weights $\alpha_{i,j}^H$ and $\alpha_{i,j}^V$ with the least squares method. The resulting optimal values of $\alpha_{i,j}^H$ and $\alpha_{i,j}^V$ in function of R are given in Fig. 21.1

Based on the experimental data from Fig. 21.1, we propose the following model for $\alpha_{i,j}^H$ and $\alpha_{i,j}^V$:

$$\widehat{\alpha}_{i,j}^H = \frac{1}{2}\frac{1}{1+R} , \tag{21.5}$$

$$\widehat{\alpha}_{i,j}^V = \frac{1}{2}\frac{R}{1+R} . \tag{21.6}$$

Note that this model fits the experimental data very well and it yields an accurate estimation of the optimal interpolation weights $\alpha_{i,j}^H$ and $\alpha_{i,j}^V$ from

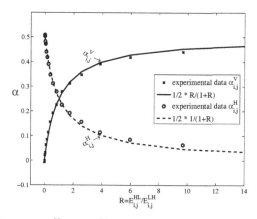

Fig. 21.1. Optimal $\alpha_{i,j}^H$ and $\alpha_{i,j}^V$ values and our approximation functions

the energy ratio R. By substituting (21.4) in (21.5) and (21.6), we obtain respectively:

$$\widehat{\alpha}_{i,j}^H = \frac{1}{2}\frac{E_{i,j}^{LH}}{E_{i,j}^{LH} + E_{i,j}^{HL}}, \tag{21.7}$$

and

$$\widehat{\alpha}_{i,j}^V = \frac{1}{2}\frac{E_{i,j}^{HL}}{E_{i,j}^{LH} + E_{i,j}^{HL}}. \tag{21.8}$$

Note that if $E_{i,j}^{LH} \gg E_{i,j}^{HL}$ (the correlation of the coefficients is much higher in the horizontal than in the vertical direction) then $\alpha_{i,j}^H \approx 1/2$ and $\alpha_{i,j}^V \approx 0$, and the lost coefficient $LL_{i,j}$ is reconstructed by horizontal interpolation as intuitively expected. Vice versa, if $E_{i,j}^{LH} \ll E_{i,j}^{HL}$, then $\alpha_{i,j}^H \approx 0$ and $\alpha_{i,j}^V \approx 1/2$, and the lost coefficient $LL_{i,j}$ is reconstructed by vertical interpolation.

If $E_{i,j}^{LH} = E_{i,j}^{HL}$, there is no preferential interpolation direction and (21.7) and (21.8) yield in this case $\alpha_{i,j}^H = 1/4$ and $\alpha_{i,j}^V = 1/4$, which is equivalent to bilinear interpolation. Note that, independent of $E_{i,j}^{LH}$ and $E_{i,j}^{HL}$, $2\alpha_{i,j}^H + 2\alpha_{i,j}^V = 1$ always holds. If $E_{i,j}^{LH} = E_{i,j}^{HL} = 0$, then we choose $\alpha_{i,j}^H = \alpha_{i,j}^V = 1/4$.

21.3 The Reconstruction of High Frequency Coefficients

High frequency coefficients are difficult to estimate because of the sparse and decorrelated representation. However, the recovery of these coefficients is of less importance, since their loss has less impact on the visual quality. This is because the coefficients are mainly zero, except near significant edges. Because high frequency content is correlated in the direction where there was only low-pass filtering and no high-pass filtering, lost LH_i and HL_i coefficients can be estimated by a one dimensional linear interpolation in the direction

Fig. 21.2. Original images: **(a)** *Lena*, **(b)** *Goldhill*, **(c)** *Tweety*, **(d)** *Nike Temple*, **(e)** *Sunset*

where there was only low-pass filtering. This technique has already proven effective in [3, 4]. As errors in the HH_i-subband are even less visible, lost HH_i coefficients are set to zero.

21.4 Results and Discussion

In this section, we compare our interpolation method with existing reconstruction methods with a similar complexity such as the bilinear concealment [3] and our previous method which is based on the globally dominant correlation direction [6]. We performed the following experiment: for five test images from Fig. 21.2 (each image has a size of 256×256), we simulated the transmission over a lossy packet network. The wavelet coefficients of each image were stored in 16 packets using the dispersive packetization strategy of [3]. By using a dispersive packetization strategy, we avoid the possibility that all neighbors of a lost coefficient are also lost, if the number of lost packets p is equal to or smaller than 4. On average, the number of lost neighbors is also minimized. If a neighboring coefficient is lost anyway, it is first approximated by averaging out its available neighbors.

After the packetization, we simulated the loss of *every combination* of p packets for $p = 1, \ldots, 4$. For $p = 1, \ldots, 4$, there are respectively 16, 120, 560 and 1820 possible combinations. The lost low frequency coefficients were repaired with three reconstruction methods: the bilinear interpolation [3], the interpolation based on the globally dominant correlation direction [6], and the proposed locally adaptive method. The lost high frequency coefficients were in all cases repaired with the same one dimensional linear filter as explained in Sect. 21.3. For each p, we calculated the average Peak Signal to Noise Ratio (PSNR) of the reconstructed images for each reconstruction method. The results of this experiment are given in Table 21.1.

If we compare the three reconstruction methods, we see that our proposed method outperforms the bilinear interpolation [3] with at least 0.5 dB and even with more than 4 dB for the *Tweety*-image. If we compare our proposed method to our previous method [6], we see that for low packet loss rates ($p = 1$), there are some images (e.g., *Sunset*, *Goldhill*) where the quality

Table 21.1. Average PSNR (dB) of the reconstructed images for $p = 1, \ldots, 4$ lost packets for the bilinear interpolation [3], our previous method based on the globally dominant correlation direction [6] and the proposed method

Average PSNR for *Lena*				Average PSNR for *Goldhill*			
p	Bilinear [3]	Global [6]	Proposed	p	Bilinear [3]	Global [6]	Proposed
1	30.58	31.10	31.43	1	33.65	33.91	33.96
2	27.42	27.86	28.22	2	30.49	30.66	30.76
3	25.49	25.87	26.26	3	28.54	28.65	28.80
4	24.05	24.39	24.79	4	27.05	27.15	27.34

Average PSNR for *Tweety*				Average PSNR for *Nike Temple*			
p	Bilinear [3]	Global [6]	Proposed	p	Bilinear [3]	Global [6]	Proposed
1	39.23	42.96	43.51	1	29.56	30.81	30.99
2	35.89	38.68	39.90	2	26.33	27.25	27.69
3	33.76	35.86	37.53	3	24.34	24.99	25.62
4	32.03	33.77	35.62	4	22.81	23.31	24.04

Average PSNR for *Sunset*			
p	Bilinear [3]	Global [6]	Proposed
1	38.80	40.48	40.45
2	35.52	36.83	37.13
3	33.45	34.49	35.03
4	31.80	32.71	33.40

of the reconstructed images is similar for both reconstruction methods. On other images (e.g., *Tweety*) we have an increase in PSNR of 0.5 dB. For high packet loss rates, our proposed method outperforms our previous interpolation technique [6] with between 0.2 and 1.85 dB.

The images used in the aforementioned experiment are all uncompressed images. In this way, we avoid mixing compression artifacts with reconstruction artifacts. In a real network application, compression is of course necessary to save bandwidth. Due to the quantization step in the compression, the wavelet coefficients are modified. As quantized coefficients may behave differently, we performed an experiment similar to the aforementioned experiment, but now only a fixed number of bit planes of the wavelet coefficients were transmitted. The number of bit planes are: 7 for *Lena*, 8 for *Goldhill*, 6 for *Tweety*, 8 for *Nike Temple* and 7 for *Sunset*. The results of this experiment are given in Table 21.2.

If no packets are lost, the PSNR of the received image is equal to the PSNR of the transmitted, compressed image. If there is packet loss, then the quality

Table 21.2. Average PSNR (dB) of the reconstructed images for $p = 1, \ldots, 4$ lost packets for the bilinear interpolation [3], our previous method based on the globally dominant correlation direction [6] and the proposed method. For these images, only a fixed number of bit planes of the wavelet coefficients were transmitted. The number of bit planes are: 7 for *Lena*, 8 for *Goldhill*, 6 for *Tweety*, 8 for *Nike Temple* and 7 for *Sunset*

Average PSNR for compressed *Lena*				Average PSNR for compressed *Goldhill*			
p	Bilinear [3]	Global [6]	Proposed	p	Bilinear [3]	Global [6]	Proposed
0	33.64	33.64	33.64	0	36.21	36.21	36.21
1	28.99	29.35	29.57	1	31.89	32.06	32.09
2	26.68	27.05	27.35	2	29.64	29.78	29.86
3	25.08	25.42	25.77	3	28.05	28.15	28.28
4	23.80	24.12	24.51	4	26.75	26.85	27.02

Average PSNR for compressed *Tweety*				Average PSNR for compressed *Nike*			
p	Bilinear [3]	Global [6]	Proposed	p	Bilinear [3]	Global [6]	Proposed
0	34.94	34.94	34.94	0	32.44	32.44	32.44
1	33.66	34.41	34.40	1	27.84	28.66	28.77
2	32.54	33.56	33.79	2	25.50	26.24	26.59
3	31.49	32.56	33.10	3	23.84	24.42	24.97
4	30.43	31.53	32.33	4	22.50	22.95	23.63

Average PSNR for compressed *Sunset*			
p	Bilinear [3]	Global [6]	Proposed
0	34.77	34.77	34.77
1	33.47	33.91	33.89
2	32.35	32.92	33.05
3	31.33	31.93	32.22
4	30.33	30.96	31.41

of the reconstructed compressed images (Table 21.2) is always lower than the reconstructed non-compressed images (Table 21.1), as there is now a quality degradation due to the compression as well as due to the reconstruction. The difference between the different reconstruction methods is roughly the same for the quantized and the non-quantized case. This means that the proposed method will also work best in case of compression.

In this experiment, we only focussed on the quantization of the wavelet coefficients and not on the source coding. Therefore, we can not give a reliable compression ratio, or a reliable size of the data packets. In future work, we will investigate a more complete compression scheme. In such a scheme it would be possible to compare the gains of different packetization schemes in terms of error concealment performance against the loss in compression performance. On the one hand, the performance of the compression algorithm is likely to deteriorate when applied on the packets of coefficients resulting from a dispersive packetization (because the coefficients within the same packet are in this case less correlated). On the other hand, a dispersive packetization facilitates and improves the error concealment (since more correctly received neighbors of a lost coefficient are available on average).

We will then also be able to determine the optimal number of decomposition levels in the wavelet transform. In the previous experiments, we used three decomposition levels ($n = 3$), which seems suitable for 256×256 images. By decreasing the number of decomposition levels, our method performs better, but on the other hand, the compression ratio will be worse. The optimal number of decomposition levels will be a trade off between the performance of the error concealment and the performance of the compression scheme.

We also visually compare the proposed method with the bilinear interpolation [3] and with our previous method which is based on the globally dominant correlation direction [6]. For these examples, we have chosen a combination of lost packets such that the resulting PSNR values are relatively close to the average PSNR values as given in Table 21.1.

In Fig. 21.3, we show *Lena* with one lost packet and we show the results after reconstruction with the three interpolation methods. Figure 21.3 (a) is the *Lena*-image after the loss of packet 5 (i.e., 6.25% of the coefficients lost). Figures 21.3 (b–d) are the images after reconstruction with respectively bilinear interpolation [3], our previous method [6], and our proposed reconstruction method. As indicated by the PSNR, the difference between Figs. 21.3 (b) and (d) is most obvious. In the bilinear interpolation, all neighboring coefficients get the same interpolation weight. This gives particularly bad results near edges. This is already better in the interpolation with a dominant correlation direction, but still, edges that do not comply with this dominant direction may still be badly interpolated. In the proposed method, the interpolation adapts to the local edge direction, yielding a better result. This is even more apparent for high packet loss rates.

In Fig. 21.4, we compare the interpolation methods for a higher packet loss rate ($p = 3$), for the *Sunset* image. Figure 21.4 (a) is the *Sunset*-image after the loss of packets 4, 7 and 15 (i.e., 18.75% of the coefficients lost). Figures 21.4 (b–d) are the images after reconstruction. The PSNR value of the image reconstructed with our proposed method is respectively more than 1.5 and 0.5 dB higher than the images reconstructed with bilinear interpolation [3], and with our previous method [6].

For $p = 4$ (i.e., 25% of the coefficients lost) we give two examples in Figs. 21.5 and 21.6, respectively for the *Tweety* and the *Nike Temple* image. Figure 21.5 (a) is the *Tweety*-image after the loss of packets 1, 8, 12, and 14. Figure 21.6 (a) is the *Nike Temple*-image after the loss of packets 2, 3, 11, and 16. Figures 21.5 (b–d) and 21.6 (b–d) are the images after reconstruction. For these examples, the difference between the three reconstruction methods

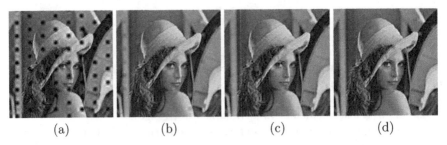

(a)   (b)   (c)   (d)

Fig. 21.3. (a) *Lena*-image after loss of packet 5. (b) Damaged *Lena*-image repaired with the bilinear interpolation [3] (PSNR = 30.88 dB). (c) Damaged *Lena*-image repaired with our previous interpolation method based on the globally dominant correlation direction [6] (PSNR = 31.59 dB). (d) Damaged *Lena*-image repaired with our proposed reconstruction method (PSNR = 31.97 dB)

(a)   (b)   (c)   (d)

Fig. 21.4. (a) *Sunset*-image after loss of packet 4, 7, and 15. (b) Damaged *Sunset*-image repaired with the bilinear interpolation [3] (PSNR = 34.07 dB). (c) Damaged *Sunset*-image repaired with our previous interpolation method based on the globally dominant correlation direction [6] (PSNR = 35.20 dB). (d) Damaged *Sunset*-image repaired with our proposed reconstruction method (PSNR = 35.74 dB)

is clearly visible: although all three methods do a good reconstruction in the smooth areas, only the proposed method succeeds in doing a satisfactory reconstruction of the significant edges. This is also reflected in the PSNR values of the reconstructed images.

To show the influence of image compression, we perform the same experiment as in Fig. 21.3, but now the wavelet coefficients are quantized to a fixed number of bit planes. Figure 21.7 (a) is a compressed *Lena*-image where only 7 bit planes of the coefficients have been transmitted. Its PSNR is 33.64 dB. Figure 21.7 (b) is the compressed *Lena*-image after the loss of packet 5 (i.e., 6.25% of the coefficients lost). Figures 21.7 (c–e) are the images after reconstruction. For each reconstruction method, the PSNR is about 1.9 dB lower for the compressed image compared with the reconstruction of the uncompressed image. Although the difference in PSNR between the reconstructed images in the compressed case is a little bit smaller than in the uncompressed case (Fig. 21.3), we can draw the same conclusions here: the proposed method

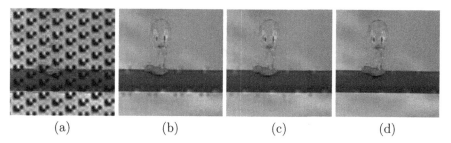

Fig. 21.5. **(a)** *Tweety*-image after loss of packet 1, 8, 12, and 14. **(b)** Damaged *Tweety*-image repaired with the bilinear interpolation [3] (PSNR = 31.71 dB). **(c)** Damaged *Tweety*-image repaired with our previous interpolation method based on the globally dominant correlation direction [6] (PSNR = 33.55 dB). **(d)** Damaged *Tweety*-image repaired with our proposed reconstruction method (PSNR = 35.50 dB)

Fig. 21.6. **(a)** *Nike Temple*-image after loss of packet 2, 3, 11, and 16. **(b)** Damaged *Nike Temple*-image repaired with the bilinear interpolation [3] (PSNR = 23.62 dB). **(c)** Damaged *Nike Temple*-image repaired with our previous interpolation method based on the globally dominant correlation direction [6] (PSNR = 24.11 dB). **(d)** Damaged *Nike Temple*-image repaired with our proposed reconstruction method (PSNR = 24.84 dB)

performs better than our previous method [6] which in turn performs better than the bilinear interpolation [3].

21.5 Conclusion

In this chapter, we presented a novel locally adaptive interpolation method for the reconstruction of lost low frequency wavelet coefficients in wavelet coded images and video. Each lost low frequency (LL^n) coefficient is interpolated from its four neighbors. The interpolation weights are estimated from the energy of the corresponding coefficients in the high frequency subbands. The ratio of the energy in the HL^n subband and of the energy in the LH^n subband are used as an indication for the magnitude of the interpolation weights.

(a)   (b)   (c)   (d)   (e)

Fig. 21.7. **(a)** Compressed *Lena*-image (PSNR = 33.64 dB). **(b)** Compressed *Lena*-image after loss of packet 5. **(c)** Damaged *Lena*-image repaired with the bilinear interpolation [3] (PSNR = 29.18 dB). **(d)** Damaged *Lena*-image repaired with our previous interpolation method based on the globally dominant correlation direction [6] (PSNR = 29.64 dB). **(e)** Damaged *Lena*-image repaired with our proposed reconstruction method (PSNR = 29.91 dB)

We have evaluated our reconstruction method by simulating the loss of every combination of one to four packets for 5 images with different content. Compared to bilinear interpolation, our method performs up to 4.3 dB better. Compared to the interpolation based on the globally dominant correlation direction, our method performs up to 0.5 dB better for low packet loss rates, and up to 1.8 dB better for high packet loss rates. These numeric results are in accordance with the visual results.

We expect that the PSNR can increase even more by taking more high frequency coefficients (e.g., from other LH^ℓ and HL^ℓ subbands) into account. A better reconstruction scheme for the high frequency coefficients is also desirable, as the simple one dimensional interpolation gives annoying artifacts for high packet loss rates.

Acknowledgments.

Aleksandra Pižurica is a postdoctoral research fellow of the FWO, Flanders, Belgium.

References

1. Wang, Y., Zhu, Q.: Error control and concealment for video communication: A review. Proceedings of the IEEE **86**(5) (May 1998) 974–997
2. Stoufs, M., Barbarien, J., Verdicchio, F., Munteanu, A., Cornelis, J., Schelkens, P.: Error protection and concealment of motion vectors in MCTF-based video coding. In: Proc. Optics East. Volume 5607., Philadelphia, PA, USA (Oct. 2004) 71–80
3. Bajić, I., Woods, J.: Domain-based multiple description coding of images and video. IEEE Trans. on Image Processing **12**(10) (Oct. 2003) 1211–1225
4. Hemami, S., Gray, R.: Subband-coded image reconstruction for lossy packet networks. IEEE Transactions on Image Processing **6**(4) (April 1997) 523–539

5. Bajić, I.: Adaptive MAP error concealment for dispersively packetized wavelet-coded images. IEEE Trans. on Image Processing **15**(5) (May 2006) 1226–1235
6. Rombaut, J., Pižurica, A., Philips, W.: Passive error concealment for wavelet coded images adapted to a directional image correlation. In: Proc. Optics East. Volume 6001., Boston, Massachusets, USA (Oct. 2005) 60010M 1–10

Content-Based Watermarking by Geometric Warping and Feature-Based Image Segmentation

Dima Pröfrock, Mathias Schlauweg, and Erika Müller

University of Rostock,
Institute of Communications Engineering,
Rostock 18119, Germany,
{dima.proefrock, mathias.schlauweg, erika.mueller}@uni-rostock.de

Summary. In this work, we present a new content-based watermarking approach that uses geometric warping to embed watermarks with high robustness to strong lossy compression. The issue of hard decisions related to content-based watermarking is discussed and it is explained why hard decisions can involve bit errors in the watermark extraction process. This work contains a solution to prevent hard decisions increasing the watermark performance. Therefore, we introduce a new feature-based image segmentation process with high robustness to lossy compression. On the basis of the segmentation, a watermark approach is proposed. Further, a secret key can be used to prevent unauthorized access to the watermark. The watermark extraction process does not need the original image. Our analyses of the watermark approach confirm the expected high robustness to strong lossy compression.

22.1 Introduction

Digital data techniques more and more replace analogue data techniques. The advantages are obviously. Digital data can be copied, edited and transferred without high efforts. However, at the same time it is very simple to make illegal copies and to manipulate digital data. Digital watermarking [1] offers contributions in protecting the authenticity of the data and the copyrights of the authors.
One property of digital watermarks is the robustness describing the possibility to extract the watermark after permitted or malicious modification of the digital data. There are watermark methods that achieve robustness to different attacks such as cropping, rotating, scaling, compression and noise ([2] gives an overview). Robustness to many attacks can be important. However, mostly these watermarks are not robust to strong lossy compression or contain only a low amount of information (watermark capacity). High robustness and a suitable capacity can be achieved using content-based watermarking

approaches [3]. Because compression algorithms try to maintain the content, the watermarks are embedded into it. The challenge of content based watermarking is to find a suitable definition of content and to solve the problem of hard decisions.

In this paper, we present a new content-based watermarking approach basing on geometric warping. Firstly, the basic idea of watermarking by geometric warping is described. Afterwards, the issue of content-based watermarking related to hard decisions is discussed. In the next section, the basic idea of content-based watermarking without hard decisions is explained. Therefore, different types of image features are introduced which are used for a feature-based segmentation and the embedding and extraction process. Finally, the results are presented including analyses of the watermark robustness to strong lossy compression.

22.2 Geometric Warping Watermarking

Generally, the performance of a watermark is defined by the watermark properties capacity, robustness and visibility. These properties depend on each other. Many watermarking methods achieve a low watermark visibility by changing the gray values of images only slightly. Some of them are SS (Spread Spectrum) [4], LSB (Least Significant Bit) [5] and DCT transformation based Quantisation Index Modulation watermarking approaches [6]. An example of Spread Spectrum watermarking is shown in Figure 22.1 a) and b). However, the approaches of changing pixels only slightly to embed the watermark compete with lossy compression algorithms. To achieve robustness to lossy compression the watermark capacity has to be reduced. Hence, it is difficulty to embed a watermark especially with robustness to strong lossy compression and a suitable capacity at the same time.

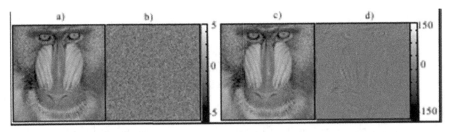

Fig. 22.1. Image "Mandrill" with SS watermarking a), corresponding difference image b), warped image "Mandrill" c) and corresponding difference image d).

- Changing the position of object borders of an image results in high difference values (see Figure 22.1 c) and d)

- Compression algorithms try to change gray values only slightly (because they are PSNR-optimized)
- Hence, compression algorithms try to maintain the position of object borders
- To embed the watermark, the position of object borders is changed by warping
- The watermark information is contained in the position of object borders and robust to strong lossy compression

A more detailed explanation can be found in [7]. Other geometric warping based watermarking approaches are given in [8] and [9].

22.3 Issue of Hard Decisions

Geometric warping based watermarking is content-based watermarking. The content can be understood as perceptually significant features in the data. The content is changed to embed the watermark with robustness to lossy compression. However, we can't embed a watermark if there is no, for the watermark method, recognizable content. Hence, to embed the watermark we have to decide what and where the content of an image is. For the latter, image segmentation is necessary.

Maes et al. propose in [8] a zero-bit [10] watermarking approach. They take the complete image to embed a watermark without any bit information. Hence, they need no segmentation. In [9], we propose a multi-bit video watermarking approach. The video is divided into two groups of blocks. In blocks which have a suitable content for watermarking and blocks without a suitable content. The watermark bits are embedded only into the suitable blocks. To extract the watermark after strong lossy compression the two groups of blocks have to be reconstructed.

The issue of hard decision in content-based watermarking is to decide whether an image region, for example a block of a block-based watermarking approach, is suitable for watermarking or not. Using this hard decision involves the probability of fail decisions after lossy compression. A fail decision can involve a watermark bit error or completely destroy the watermark.

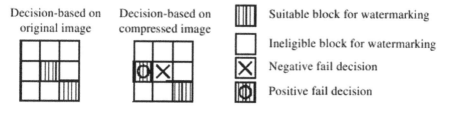

Fig. 22.2. Principle of failed hard decision.

A possibility to solve the problem of fail decision is to use error correction codes. Common error correction is designed to correct only substitution errors. The fail decisions involve insertions (negative fail decisions) and deletions (positive fail decisions). In [11], Schlauweg et al presented a coding technique that is also able to correct insertion and deletion errors.

In [9], we use a pre-distortion of the blocks to prevent fail decisions. A block-feature combined with a threshold t is used to decide whether a block is suitable for watermarking or not. To reduce the probability of fail decisions, blocks with a value of the feature near the threshold are pre-distorted. The process changes the blocks in a way that the new value of the feature has a higher distance to the threshold t. After pre-distortion there is a gap in the feature-value distribution.

The problem of hard decision can be solved on several ways. However, every solution has disadvantages. Whereas the use of error correction codes reduces the watermark capacity, creating a gap requires a pre-distortion and increases the watermark visibility. The advantage of a content-based watermarking approach without hard decisions is an increased performance of the watermark. In the following sections, we propose a multi-bit geometric warping based watermarking method without hard decisions.

22.4 A Geometric Warping Watermark Approach Without Hard Decisions

22.4.1 Basic Idea

Content-based watermarking uses features which describe the content of an image. For content-based watermarking without hard decision we propose to use two different types of features.

The first feature describes the type of the content (Locator-Feature). For example, the Locator-Feature could describe the strength of the edges inside a block. This feature is used to find content which is suitable for watermarking. The second feature describes the content (Carrier-Feature). For example, the Carrier -Feature could describe the position of the edge inside a block.

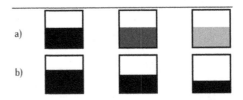

Fig. 22.3. Example with a) different content type (Locator-Feature - edge strength) and b) different content (Carrier-Feature - edge position).

To embed the watermark, the Locator-Feature is used to find content which has a suitable Carrier-Feature. The embedding process changes the content without changing the type of content. Starting from these assumptions we can create a content-based geometric warping watermarking approach without hard decisions.

We propose to use a Locator-Feature-based segmentation process to get only suitable blocks respectively segments for watermarking. A hard decision is not necessary. The image can be divided into segments even if the content is uniform or non-uniform (see Figure 22.4). The Carrier-Feature of each segment is used to carry one watermark bit. The segmentation can be reconstructed after strong lossy compression if both features are robust to strong lossy compression.

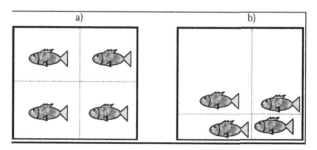

Fig. 22.4. Simplified principle of Locator-Feature based image segmentation. Each of the four segments contains the same "amount" of content. It doesn't matter if the content is uniform a) or non-uniform b).

22.4.2 Locator- and Carrier Feature

In [9], we propose the NCG (Normed Centre of Gravity). The NCG is a block-based statistic. It describes the strength and position of the gravity centre of a block in a block border independent way. To compute the NCG of a block with size $n \times n$, the mean values of all columns and rows yielding the vectors m_x and m_y are used (see Figure 22.5). Both vectors are used to compute the 2-dimensional vector v_k ($k = x$ or y):

$$\underline{v}_k = \begin{pmatrix} \sum_{i=1}^{n} \underline{m}_k(i) \cdot \cos\left(\frac{\pi}{n} + \left((i-1) \cdot \left(\frac{2 \cdot \pi}{n}\right)\right)\right) \\ \sum_{i=1}^{n} \underline{m}_k(i) \cdot \sin\left(\frac{\pi}{n} + \left((i-1) \cdot \left(\frac{2 \cdot \pi}{n}\right)\right)\right) \end{pmatrix} \quad (22.1)$$

For both vectors, angles Θ_k are computed. These values are used to compute the x,y-coordinates of the NCG:

$$x = \frac{n \cdot \Theta_x}{2 \cdot \pi} \qquad y = \frac{n \cdot \Theta_y}{2 \cdot \pi} \quad (22.2)$$

Locator-Feature:

The strength L of the NCG is used as Locator-Feature. It bases on the vector length L_k of v_k:

$$L = \sqrt{L_x^2 + L_y^2} \qquad (22.3)$$

Carrier-Feature:

The NCG x,y-coordiantes are mapped on a self adapting quantization lattice. The resulting value s is very robust to lossy compression. For a detailed description of this process, see [9]. This value s is used as Carrier-Feature.

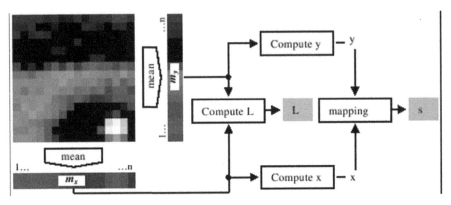

Fig. 22.5. Overview of the Locator- and Carrier-Feature calculation scheme.

To realize a Locator-Feature based image segmentation, the NCG has to be computed for each pixel. Therefore, the pixels surrounding the current pixels are used to build a block. The NCG of the block represents the NCG of the current pixel.

Each pixel delivers a Locator-Feature and a Carrier-Feature. The results are the Locator-Feature matrix LFM and the Carrier-Feature matrix CFM. For example, see Figure 22.6.

Original image   Locator-Feature matrix   Carrier-Feature matrix

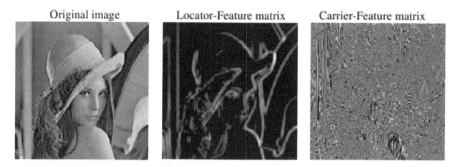

Fig. 22.6. Example of the NCG Locator- and Carrier-Feature matrices.

22.4.3 Locator-Feature-Based Image Segmentation

The aim of Locator-Feature-based image segmentation is a uniform qualifica-
tion of each segment to carry a watermark bit. It is necessary to reconstruct
the segments even after strong lossy compression. Hence, the segments respectively the segmentation process has to be robust to strong lossy compression.
A higher value L (Locator-Feature) yields a higher robustness of the value
s (Carrier-Feature). Hence, one segment needs either many pixels with low
values L or only some pixels with high values L to carry a watermark bit
with the same robustness to lossy compression. Because of this, we propose to
divide the image into segments where the sum of the LFM elements in each
segment is the same. Therefore, following algorithm is used:

1. The columns of LFM are averaged.
2. The resulting vector is divided into a segments whereby the sum of the
   vector elements in each segment is equal.
3. Each vector segment represents a set of columns in LFM.
4. Each set of columns is divided into b segments by averaging the rows
   and dividing the resulting vector into b segments whereby the sum of the
   vector elements in each segment is equal.
5. The results are $a \cdot b$ segments where the sum of the LFM elements in each
   segment is the same (see Figure 22.7).

The robustness of the segments is analyzed by using JPEG and JPEG2000
(JasPer-Codec). As shown in Figure 22.8, the mean error of the segmentation
is relatively low also after strong lossy compression. The maximal mean error
of about 10% means that 90% of a segment respectively 90% of the location
of the Carrier-Feature can be reconstructed.

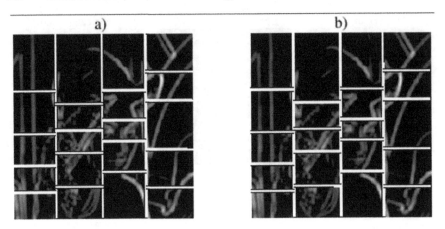

Fig. 22.7. Example of Locator-Feature-based segmentation with a) segments of original image "Lena" and b) segmentation after strong lossy compression (JPEG with quality factor 15).

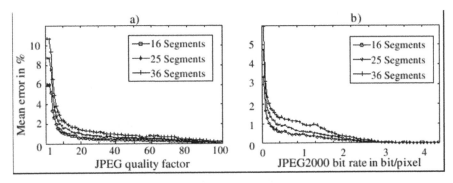

Fig. 22.8. Robustness of the segments to a) JPEG and b) JPEG2000 compression.

22.5 Watermark Embedding and Extracting

To embed the watermark bit, matrix LFM, CFM and a pseudo random binary pattern BP are used. The binary pattern has the same size as matrices LFM and CFM and can be created by a known algorithm or a secret key. Hence, the watermark can be protected against unauthorized access. For example, see Figure 22.9. Matrix LFM is normalized to a value range between 0 and 1 and element wise multiplied with matrix CFM (Figure 22.9 d)).

$$LCM = LFM_{normalized} \cdot CFM \qquad (22.4)$$

The resulting matrix LCM is very robust to lossy compression. To create a relationship between the elements of matrix LCM and their spatial positions matrix LCM is element wise multiplied with matrix BP (Figure 22.9 e)). Result is matrix $LCBM$.

$$LCBM = LCM \cdot BP \qquad (22.5)$$

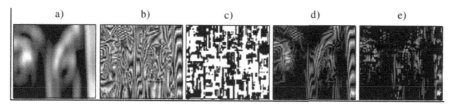

Fig. 22.9. Segment of a) Locator-Feature matrix, b) Carrier-Feature matrix, c) binary pattern, robust matrix LCM d) and of $LCBM$ e).

The elements of LCM are maintained in $LCBM$ where the equivalent elements of BP have the value one. The elements of LCM are set to zero in $LCBM$ where the equivalent elements of BP have the value zero. The relationship between segment k of $LCBM$ and segment k of LCM is the scalar SR_k.

$$SR_k = \frac{\sum\limits_{i=1}^{m}\sum\limits_{j=1}^{n} lcbm_{k_{i,j}}}{\sum\limits_{i=1}^{m}\sum\limits_{j=1}^{n} lcm_{k_{i,j}}} \qquad (22.6)$$

SR_k is the basis of the embedding process and has a value range between 0 and 1. To embed a bit value '0' respectively '1' the image is changed by geometric warping so that $0 <= SR_k < 0.5$ respectively $0.5 <= SR_k <= 1$. Analysis of the SR_k robustness to lossy compression shows that SR_k is suitable to carry the watermark. As shown in Figure 22.10 the MAE (Mean Absolute Error) is very low. This error analysis considers already the error caused by the Locator-Feature based segmentation process.

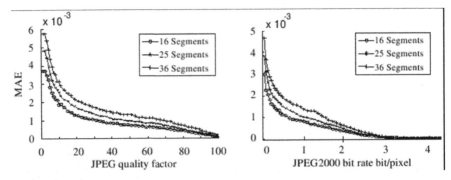

Fig. 22.10. Robustness of SR_k to a) JPEG and b) JPEG2000 compression.

To embed the watermark bits, the image has to be warped. Therefore, a warping matrix is computed in a way that the new resulting SR_k have the wanted values. For example, see Figure 22.11.

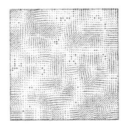

Fig. 22.11. Example of warping matrix gained by factor 10 for a better visualization.

The embedding process is computationally expensive. The largest amount of computing power is used to get the warping matrix. However, the watermarking extraction process doesn't need the warping matrix. Hence, extracting the watermark bits needs less computing power. The extraction process requires the matrices LFM, CFM and BP to realize the segmentation process and to compute SR for each segment. These values can be computed directly using the watermarked image. The original image is not needed. The watermark bit values '0' or '1' can be directly computed using the single SR_k.

22.6 Results

The watermarking approach was tested for different gray scaled images (Miscelaneous database) with the size of 512x512 pixels. A maximal warping strength of one was used. Hence, the spatial position of a pixel is moved less than one by the warping process. The number of embedded bits respectively the number of segments is 16, 25 and 36. An example can be seen in

Figure 22.12.

a)   b)

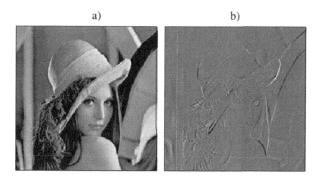

Fig. 22.12. Watermarked image "Lena" a) and difference image between water-marked image and original b). The warping strength is gained by factor 3 for a better visualization.

The results of the robustness analysis are shown in Figure 22.13. The ro-bustness analysis considers the robustness of the segmentation process and the robustness of SR_k. As expected, the watermark is robust to strong lossy compression. On a JPEG compression with quality factor 1, 99.45% of 16 embedded bits can be correct extracted. On a JPEG2000 compression with a resulting bit rate of 0.08 Bit/Pixel, 99.96% of 16 embedded bits can be correct extracted.

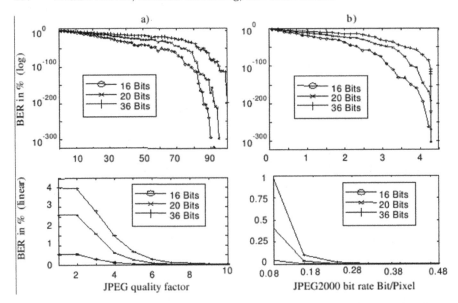

Fig. 22.13. Bit Error Rate (BER) of the embedded watermark bits after lossy compression with a) JPEG compression and b) JPEG2000 compression.

22.7 Conclusion

In this work, we propose a content-based watermarking approach basing on geometric warping. The suitability of geometric warping based watermarking approaches to achieve high watermark robustness to lossy compression is explained. The issue of hard decisions related to content-based watermarking is discussed. We propose a solution to prevent hard decisions increasing the watermark efficiency. Therefore, we introduce two new types of features and a feature-based segmentation process. The robustness of the segmentation process is analysed and presented. The proposed watermarking method offers the possibility to protect the watermark against unauthorized access. The watermark extraction process does not need the original image. Analyses confirm the expected high robustness to lossy compression.

References

1. F. Perez-Gonzales, J. R. Hernandez: A tutorial on digital watermarking. Proc. of International Carnahan Conference on Security Technology (1999) 286–292
2. S. Voloshynovski, S. Pereira, T. Pun, J. J. Eggers, J. K. Su: Attacks on Digital Watermarks: Classification, Estimation-based Attacks and Benchmarks In IEEE Communications Magazine, **39** (2001) 118–126
3. M. Kutter, S. K. Bhattacharjee, T. Ebrahimi: Towards Second Generation Watermarking Schemes. Proc. of ICIP **1** (1999) 320–323

4. I. J. Cox, J. Kilian, T. Leigthon and T. Shamoon: Secure spread spectrum watermarking for images audio and video. Proc. of ICIP **3** (1996) 243–246
5. C. Rajaratnam, N. Memon: Analysis of LSB-based Image Steganography techniques. Proc. of ICIP **3** (2001) 1019-1022
6. B. Chen, G. W. Wornell: Quantization index modulation methods for digital watermarking and information embedding of multimedia. Journal of VLSI Signal Processing Systems for Signal, Image, and Video Technology, Special Issue on Multimedia Signal Processing **27** (2001) 7–33
7. D. Pröfrock, M. Schlauweg, E. Müller: A new uncompressed-domain Videowatermarking approach robust to H.264/AVC compression. Proc. of Signal Processing, Pattern Recognition and Applications (SPPRA) ISSN 1482-7921 (2006) 99–104
8. M. J. J. J. B. Maes, C. W. A. M. van Overveld: Digital watermarking by geometric warping. Proc. of ICIP **2** (1998) 424–426
9. D. Pröfrock, M. Schlauweg, E. Müller: Video Watermarking by Using Geometric Warping Without Visible Artifacts. Proc. of Information Hiding (8th IH 2006) (2006)
10. I. J. Cox, M. L. Miller, J. A. Bloom: Digital Watermarking. ISBN 1558607145 (2001)
11. M. Schlauweg, D. Pröfrock,E. Müller: Soft Feature-Based Watermark Decoding with Insertion/Deletion Correction. Proc. of Information Hiding (9th IH 2007) (2007)

Hardware Based Steganalysis

Kang Sun, Xuezeng Pan, Jimin Wang, and Lingdi Ping

College of Computer Science and Technology, Zhejiang University, Hangzhou, 310027, China.
swankong@126.com

Summary. Steganalysis is the reverse process of steganography. The goal of steganalysis is to detect, as reliably as possible, the presence of hidden data. Software-based steganolytic systems often fail to keep up with high-speed network throughputs. In this chapter, we present the design of a system that automatically detects steg-information in real-time. In this system, RS steganalytic algorithm is parallel implemented with a three-stage pipeline based on FPGA. Experiment results show that this system can achieve very high throughputs (2.5Gbps) and deal with a far larger amount of traffic than software-based approaches.

Key words: steganalysis, FPGA, RS, steganography, reconfigurable computing

23.1 Introduction

Steganography is the art of secret communication. We can use digital images, videos, audios, and other computer files that contain irrelevant or redundant information as covers or carriers to hide secret messages [1]. Steganography has made positive contributions to the field of information security. Many steganographic software and watermarking algorithms can be downloaded freely from the Internet. People might use these tools to communicate secretly with each other. However, it can also be employed by criminals - terrorists can use steganography to transmit secret messages on internet or launch terrorist attacks.

Steganalysis is the reverse process of steganography. The goal of steganalysis is to detect, as reliably as possible, the presence of hidden data. On-line real time detection is an effective way to detect hidden data transmitted on internet. But due to huge network traffic, the throughput of existing solutions cannot satisfy the requirements of on-line detection.

In this chapter, we place a strong focus on high throughput implementation of steganalytic algorithm. The architecture of an FPGA based LSB steganography detector is introduced. It uses RS [2] steganalytic algorithm to detect information hidden in color or gray-scale images. The potential advantages of using FPGA to implement steganalytic algorithm include:

Algorithm Agility This term refers to the switching of steganalytic algorithms during operation. In a real network environment, the hidden messages may be embedded in the media by various steganographic programs. It requires the detector can deal with multiple steganographic technologies, and future extensions should be possible. Whereas algorithm agility is costly with traditional hardware, FPGAs can be reprogrammed on-the-fly. And it is perceivable that fielded devices are upgraded with a new steganalytic algorithm which does not exist (or was not standardized) at design time.

Throughput Although typically slower than ASIC implementations, FPGA implementations have the potential of running substantially faster than software implementations.

Cost Efficiency The time and costs for developing an FPGA implementation of a given algorithm are much lower than for an ASIC implementation. (However, for high-volume applications, ASIC solutions usually become the more cost-efficient choice.)

23.1.1 Our Contribution

The primary contribution of our work has been to first implement RS steganalytic algorithm on reconfigurable hardware. In order to achieve high throughput, we propose a completely new reconfigurable staganography detector architecture in which the RS steganalytic algorithm is implemented in full parallel mode. Some critical operations in the algorithm are carried out by lookup table operation to accelerate the processing speed. To the best of our knowledge, there's no hardware implementation of steganalytic algorithm up to now.

23.1.2 Organization of the Chapter

In the rest of this chapter, Section 23.2 provides the background of our work in term of an introduction to the RS steganalytic algorithm and recent previous work on acceleration of LSB steganography detection based on various technologies. Section 23.3 contain the details of the system architecture. Section 23.4 presents the implementation results. At last, in Section 23.5, we conclude the whole chapter.

23.2 Background and Related Work

23.2.1 RS Steganalytic Algorithm

RS steganalytic algorithm is proposed by Fridrich, et al [2]-[4]. The stego-detection method starts with dividing the image into disjoint groups of n

adjacent pixels $(x_1, \ldots, x_n)$. For example, we can choose groups of $n = 4$ consecutive pixels in a row. A discrimination function f is defined and each pixel group G is assigned a real number $f(x_1, \ldots, x_n) \in R$. The purpose of the discrimination function is to quantify the smoothness or "regularity" of the group of pixels G. The noisier the group of pixels $G = (x_1, \ldots, x_n)$ is, the larger the value of the discrimination function becomes. So, the function f can be defined as the 'variation' of the group of pixels G:

$$f(x_1, \ldots, x_n) = \sum_{i=1}^{n-1} | x_{i+1} - x_i | . \tag{23.1}$$

Then an invertible operation F on pixel groups called "flipping" is defined as a permutation of gray levels that consists of two cycles. Thus, $F(F(x)) = x$ for each pixel x. The permutation F_1: $0 \leftrightarrow 1, 2 \leftrightarrow 3, \ldots, 254 \leftrightarrow 255$ corresponds to flipping (negating) the LSB of each gray level. And the permutation F_{-1}: $-1 \leftrightarrow 0, 1 \leftrightarrow 2, 3 \leftrightarrow 4, \ldots, 253 \leftrightarrow 254, 255 \leftrightarrow 256$, or defined as

$$F_{-1}(x) = F_1(x+1) - 1, \tag{23.2}$$

is called shifted LSB flipping. For completeness, another flipping operation called identity permutation is defined as F_0: $F_0(x) = x$. Using the discrimination function f and the flipping operation F, three types of pixel groups R, S and U are defined:

$$\mathbf{R}egular Groups : G \in R \Leftrightarrow f(F(G)) > F(G)$$

$$\mathbf{S}ingular Groups : G \in S \Leftrightarrow f(F(G)) < F(G)$$

$$\mathbf{U}nusable Groups : G \in U \Leftrightarrow f(F(G)) = F(G),$$

where $F(G) = (F(x_1), \ldots, F(x_n))$. The assignment of flipping to pixels can be captured with a mask M, which is an n tuple with values -1, 0, and 1. The flipped group $F_M(G)$ is defined as $(F_{M(1)}(x_1), F_{M(2)}(x_2), \ldots, F_{M(n)}(x_n))$. The number of regular and singular groups for mask M are denoted as R_M and S_M respectively. Similarly, the number of regular and singular groups of negative mask $-M$ are denoted as R_{-M} and S_{-M}. In a typical image, the expected value of R_M is equal to that of R_{-M}, and the same is true for S_M and S_{-M}:

$$R_M \cong R_{-M}, S_M \cong S_{-M} \tag{23.3}$$

But after randomizing the LSB plane, the above equations are violated. The principle of the RS method is to estimate the four values: R_M, R_{-M}, S_M and S_{-M}.

23.2.2 Previous Work in Steganalytic Algorithm Implementation

In the past several years, great achievements have been made in research of steanalysis. However, most of the researchers focused on the detection accuracy and gave little thought to the throughput issue. Lang, Xia and Zhi, et al [5] implemented several steganalytic algorithms and analyzed their performance. In their paper, the confidence interval of RS method was presented, but no data about the algorithm throughput was given. Zhang and Ping [6] implemented RS and Difference Histogram (DH) based steganalytic algorithms by Visual C++ on an Intel Pentium III 600MHz machine. The throughput rate of RS and DH algorithms were about 550KB/s and 4MB/s, respectively. To achieve higher processing throughput, computer clusters have been proposed to offload the workload of a single computer. Andrew [7] employed a distributed network of computers to evaluate "Pairs" and "RS" steganalytic algorithms and compared their performance. The cost of the computer cluster remains high, however, because it requires multiple processors, a distribute network and a clustered management system. In this chapter, we present a solution to implement steganalysis completely in reconfigurable hardware. There's so far no literature about this method.

23.3 Proposed FPGA Based RS LSB Steganography Detector

23.3.1 System Architecture

The architecture of the steganography detector is shown in Fig. 1. The whole detector consists of four components: address generator (AG), block memory, reconfigurable steganography detect engine (RSDE), and main microcontroller (MCU). AG is responsible for calculating the addresses which are used to access the block memory. It supports several data scan patterns to organize the pixel group G which is defined in RS algorithm (See Section 23.2.1). RSDE is the computing kernel of this detector, where RS algorithm is implemented with a 3-stage pipeline. The block memory is a dual-port SRAM which consists of two memory banks. It is used for storage of image pixels, and lookup tables. The data width of block memory is 16 bits. The MCU is an embedded microprocessor, such as MicroBlaze in Xilinx FPGA, which is in charge of harmonizing and commanding different parts to work.

23.3.2 Address Generator

The reconfigurable steganography detect engine (RSDE) is driven by data stream instead of instruction stream. Address generator (AG) is responsible for organizing data and feeding them to RSDE. The pixels of a color image that are stored in the block memory form a 3-dimensional data space. At run time, an address stream is generated by AG and the accessed data are passed from the block memory to RSDE. This principle is derived from the fact that in RS algorithm, the image must firstly be divided into several pixel groups, but the division mode is not fixed, so several addressing modes are designed

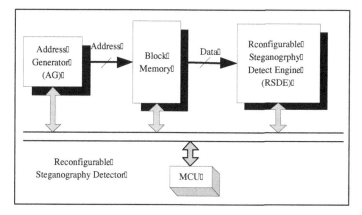

Fig. 23.1. System architecture of the proposed LSB steganography detector.

for various needs. In our design, the number of pixels per group is 4 and the data bus width is 32 bits. Two addressing modes have been so far supported by the AG - row-based scan mode and 2x2 array scan mode which are shown in Fig. 2.

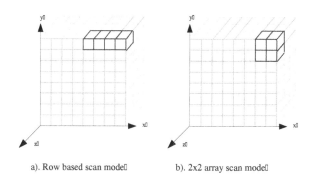

a). Row based scan mode   b). 2x2 array scan mode

Fig. 23.2. Two scan modes supported by AG.

We have limited our first implementation to an AG which can handle regular mappings in 3 dimensions. The AG can also handle 2-dimensional mappings which appear in gray scale images. To generate an address stream according to a 1,2 and 3- dimensional data space, we defined a parameters set which is called address parameter vector (APV). It consists of five parameters:

- B: Base address - the first address which is sent to the memory.
- N_x: The number of addresses in each line in the data space.
- N_y: The number of lines in the y-direction in the data space.
- N_z: The number of planes in the z-direction, if the set comprises 3 dimensions. For color images, it is a constant of 3.

- M: Address generation mode corresponding to the data scan mode shown in Fig. 23.2.

Fig. 23.3. shows the block diagram of the AG. The five registers hold the APV for address calculation. At runtime, Address Calculation Unit (ACU) is responsible for calculating the address according to the scan mode. AG gives two outputs - Addr A and Addr B, which are the addresses for the two memory banks, respectively. In row-based scan mode, AG firstly generates an address stream for every four pixels on R plane, then followed by G and B planes. While in 2x2 scan mode, the pixels are stored in the two memory banks according to their line number (odd or even) and AG firstly generate an address stream for every two pixels on R plane for both memory bank A and B simultaneously, then followed by G and B planes. Currently, in our design, the width of address bus is 22 bits.

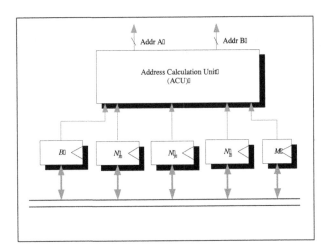

Fig. 23.3. Block diagram of address generator

23.3.3 Reconfigurable Steganography Detect Engine

Fig. 23.4 depicts the architecture of RSDE. The input pixel groups first pass through the flipping module. In RS algorithm, there are three flipping operations which are defined as F_1, F_0 and F_{-1}. F_0 is an identity permutation which does not change the pixel value. F_1 can be realized by exclusive-ORing the least significant bit of the data with 1. If F_{-1} is implemented in terms of equation (23.2), it will require two add (subtract) and one exclusive-OR operations. This solution seems to be more time-consuming compared with F_1 which needs only one exclusive-OR operation. So we use table lookup operation instead of calculation of equation (23.2) to realize F_{-1}. When implementing the circuit on a Xilinx FPGA, the table is implemented by configuring dual-port, on-chip block RAM as an array of memory locations. Each of the memories can afford two read operations every clock cycle. We built two identical tables for the four shifted flipping operations (F_{-1} flipping) that perform simultaneously, each table contains 256 bytes.

To achieve high performance, the whole circuit is pipelined with three stages: flipping operation stage, discrimination function stage and RS value calculation stage. They correspond to the stage 1, 2, 3 in Fig. 23.4, respectively. In stage 1, all three flipping operations are carried out in full parallel mode. The results are stored in registers and transmitted to stage 2 in next clock cycle. Stage 2 performs discrimination value calculation. The multiplexers select the output data generated by flipping module and send them to the discrimination function module. The selection of output depends on the mask signal M, which is an n-tuple with values 1, 0, -1. In our design, the value of n is 4 and we use two bits to encode each member in the tuple: '00' and '11' represent the value of '1' and '-1' respectively, and both '01' and '10' represent the value '0' in M. With this encode mode, we can easily get -M from M by an inverter (see Fig. 23.4.). Fig. 23.5 shows the structure of discrimination function module. It consists of one adder, three subtracters and several multiplexers and comparators. This module calculates the discrimination function value of the pixels according to equation (23.1). The RS statistic values are computed in stage 3. In Fig. 23.4, r_M, r_{-M}, s_M and s_{-M} are four counters that calculate and store the value of R_M, R_{-M}, S_M and S_{-M}, respectively.

23.4 Implementation Results

We used Xilinx VirtexII XC2V3000 which contains 3584 CLBs (Configurable Logic Blocks) as our target device [8]. The steganography detecting system was designed by Verilog HDL. ModelSim 6.0 SE by Mentor Graphics was used to perform behavioral and timing simulations for the whole system. The simulations verified both the functionality and the ability to operate at the designated clock frequencies for the implementations. The synthesis tool was Xilinx ISE 7.1. To show the performance, we list the synthesis results of RSDE (not including the microprocessor) in Table 23.1. Throughput is calculated as:

$$Throughput = 32 Bits * Clock Frequency \qquad (23.4)$$

The number of CLBs required as well as the maximum operating frequency for the implementation was obtained from the Xilinx report files. Note that the Xilinx tools assume the absolute worst possible operating conditions highest possible operating temperature, lowest possible supply voltage, and worst-case fabrication tolerance for the speed grade of the FPGA [9]. As a result, it is common for actual implementations to achieve slightly better performance results than those specified in the Xilinx report files.

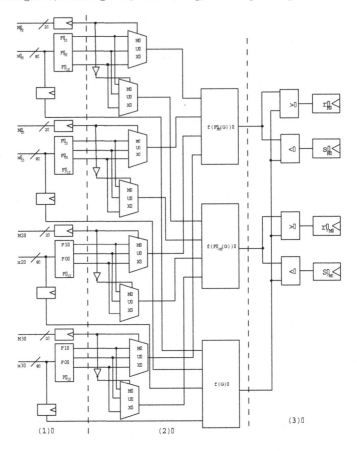

Fig. 23.4. The architecture of RSDE.

We chose the 1096 pictures in CorelDRAW 10 distribution as our experimental image library. A series of stego-inforamtioin was created from the origi-

Table 23.1. Synthesis results of RSDE.

Compile Time	2.63s
Number of Slices	263
Number of Slice Flip Flops	211
Number of 4 input LUTs	499
Number of bonded IOBs	171
Number of BRAMs	2
Clock Frequency	78.775MHZ
Throughput	2.52Gbits/s

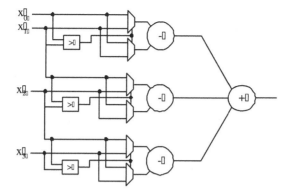

Fig. 23.5. The structure of discrimination function module.

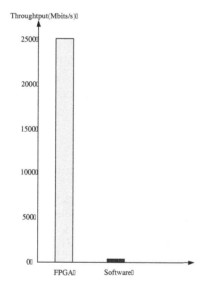

Fig. 23.6. Performance comparison between FPGA and software method.

nal image by randomizing the LSBs of 0 - 95% pixels in 10% - 15% increments. Groups of 2x2 pixels with the mask [1, 0, 0, 1] were used in our experiment. The simulation results were consistent with [2]-[3]. To compare performance, we also implemented RS algorithm in software. The program was compiled by Microsoft Visual C++ 6.0. The experiment platform is an AMD Athlon XP running Microsoft Windows XP professional. The CPU clock rate is 1.5GHz and the main memory is 512MB. The average throughput is 18.94Mb/s. Fig. 23.6 shows the comparison between these two methods (FPGA and software). The results show that the performance of FPGA solution is far higher (about 130 times) than software solution.

23.5 Conclusions

In this chapter, the design of a system for LSB steganography detection is presented. In this system, RS steganalysis algorithm is parallel implemented with a three-stage pipeline based on Xilinx Virtex II FPGA. Simulation and synthesis results show that the system is capable of achieving very high throughputs (about 2.5Gbps). Since we exploit the parallelism afforded by hardware, the system is able to deal with a far larger amount of traffic than software based approaches.

As future work, the hardware detector can be enhanced further to achieve processing of data with even higher throughput. And we also plan to integrate more steganolytic algorithms into our detecting system to improve the system ability.

Acknowledgements.

This work was supported by Natural Science Foundation of Zhejiang Province, China (Grant No.Y105355), special project of Zhejiang High-Tech Development Plan (Grant No.2006C11105), and project of Hangzhou Government Office of Research and Industrialization (Grant No.20061331E16).

References

1. Cole E (2003). Hiding in Plain Sight: Steganography and the Art of Covert Communication. Wiley, ISBN: 0471444499.
2. Fridrich J, Goljan M, Du R (2001). Detecting LSB Steganography in Color and Gray-scaleImages. IEEE Multimedia Magazine. 8(4): 22-28.
3. Fridrich J, Goljan M, Du R (2001). Reliable Detection of LSB Steganography in Color and Grayscale Images. Proceedings ACM, Special Session on Multimedia Security and Watermarking. Ottawa, USA : 27-30.
4. Fridrich J, Goljan M (2002). Practical Steganalysis of Digital ImagesCState of the Art. Proceedings of the SPIE, Security and Watermarking of Multimedia Contents IV. San Jose, USA. 4675: 1-13.
5. Lang R L, Xia Y, Zhi Y, Dai G Z (2004). Analysis and Evaluation of Several Typical Steganalysis Algorithms. Journal of Image and Graphics, 9(2): 249-256. (in Chinese)
6. Zhang T, Ping X J (2004). Reliable Detection of Spatial LSB Steganography Based on Difference Histogram. Journal of Software, 15(1): 151-158. (in Chinese)
7. Andrew D K (2004). Quantitative Evaluation of Pairs and RS Steganalysis. Proceedings of SPIE, Security, Steganography, and Watermarking of Multimedia Contents VI. 5306: 83-97.
8. Xilinx (2005). Virtex-II Platform FPGAs: Complete Data Sheet. v3.4 http://direct.xilinx.com/bvdocs/publications/ds031.pdf
9. Blum T, Paar C (2001). High-radix Montgomery Modular Exponentiation on Reconfigurable Hardware. IEEE Transactions on Computers, 50(7): 759-764.

Esophageal speech enhancement using source synthesis and formant patterns modification

Rym Haj Ali[1] and Sofia Ben Jebara[2]

[1] Ecole Superieure des Communications de Tunis
  rym.elhadjali@gmail.com
[2] Ecole Superieure des Communications de Tunis
  sofia.benjebara@supcom.rnu.tn

Summary. *This paper deals with esophageal speech, which is a voice of substitution used by alaryngeal persons in order to be able to communicate with others. This voice, characterized by a low intensity and poor intelligibility, is hard to understand. In this paper, we propose ideas to enhance this kind of voice. More precisely, we enhance the source excitation signal and the formant structure of the speech vocal track. We modify pitch values by those of natural speech and we replace the source by a synthetic one based on LF model. We also enhance the formant structure by enlarging formant bandwidth and we amplify their amplitudes without increasing the background noise. Then, we englobe all modification in the same scheme including all improvements.*

Key words: Esophageal speech, Enhancement, Pitch extraction, Formant pattern.

24.1 Introduction

For some persons suffering from the cancer of larynx, it is necessary to proceed the larynx ablation so that they will loose their voice and and will not be able to speak. The medical staff invented some elementary prothesis to permit to these persons to communicate with others. The esophageal speech, an alternative solution having the advantage of not using any artificial machine, is the most naturel approach. It consist on injecting air into the esophagus extremity. The vibration, usually created in the vocal cords during the natural speech, is now created in the esophagus extremity. This esophageal voice is very noisy and has low intensity and poor intelligibility.

This paper deals with esophageal voice to make it more comprehensible and less noisy. In literature, there are some works which deal with this problem. we relate for example works using pattern recognition techniques [1], speech based filtering [2, 3],...

In this paper, we present a different approach to make the esophageal voice more understood and enjoyable with more intensity. According to the natural speech model, we will first propose to enhance separately the two main signals characterizing any speech which are the excitation signal and the vocal track signal, defined by its formantic structure. Next, we combine the improvement in different manners in order to obtain a global enhanced speech signal.

This paper is organized as follows. In section 2, we will give the proposed solution flowchart and its main ideas. In section 3 (resp. section 4), we will develop the approach for source excitation (resp. vocal track) signal improvement. In section 5, we will present the result we get.

24.2 Synoptic of the solution

Our method in analyzing esophageal speech is based on straightforward source and formant analysis/synthesis method. In fact, before enhancing the esophageal speech, it is necessary to extract its essential components: the excitation signal and the vocal track signal and then modify their characteristics. The basic scheme is shown in figure 24.1.

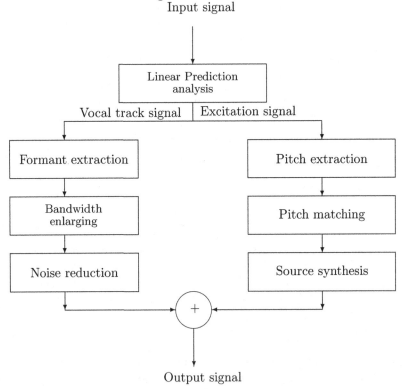

Fig. 24.1. Synoptic of the proposed solution

The input signal is sampled at 16 Khz and decomposed into 30 ms frames. Each frame is divided in two main signals: the excitation signal and the vocal track signal through linear predicio analysis of order N=10. Each subsignal is treated separately. Each excitation frame is also classified as voiced or unvoiced frames in order to calculate pitch value. The unvoiced frame is untouchable and will be synthesized at the end. An auto-correlation based pitch analysis is performed. Pitch values are adjusted and then used to generate the synthetic excitation signal. On the other hand, the vocal track signal is analyzed in order to calculate formant frenquencies, bandwitdhs and energy. Formant bandwidths are enlarged by a perceptual filtering which will be described later. As a consequence, the hole signal is amplified. The last step is to merge the enhanced signals to create a enhanced esophageal speech.

24.3 The Source enhancement

The speech takes form at the excitation source. So, the speech enhancement begins by the excitation signal enhancement. For unvoiced frames,we propose the use of the classical random white gaussien source. In case of voiced frames, we propose the following procedure based on pitch calculas, pitch matching and Liljencrants-Fants source synthesis.

24.3.1 Pitch extraction

Since esophageal speech source is different from natural speech source, it is obvious that any technique of natural pitch extraction does not work, that's why we adapt a natural speech classical technique to esophageal speech. In this paper, we propose to adapt the Modified Auto-Correlation Method (MACM), creating the Esophageal Voice-Modified Auto-Correlation Method (EV-MACM). To attend this gate, we add three criteria for the decision of voiced/unvoiced decision. They are the zero crossing rate, the frame energy and the Bindex. It is the ratio between high frequency energy and low frequency energy. In case of voiced frames, there is a pic of energy in the low frequency. So that the Bindex is high.

The flowchart of the EV-MACM is shown in the figure24.2. After the calculas of the three criteria, we decide wether the frame is voiced or not.

The figure 24.3 illustrates and compares pitch values of natural speech (a) and esophageal speech (b) for the french sequence "aoa". We observe that the esophageal pitch is lower than the natural one. In fact, the esophageal pitch is generally between 50 Hz and 150 Hz (24.3) whereas natural pitch can reach 300 Hz. So, a first step of enhancement is to replace the esophageal pitch with the naturel pitch.

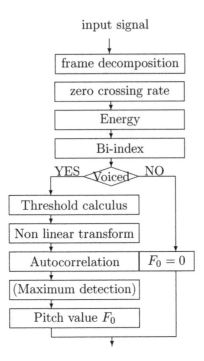

Fig. 24.2. EV-MACM flowchart.

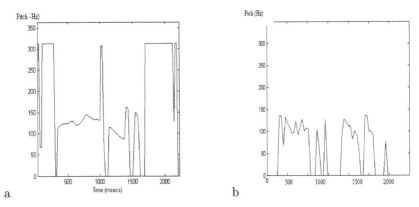

Fig. 24.3. Comparison between pitch values for natural (a) and esophageal (b) speech for the sequence "aoa"

24.3.2 Natural pitch and esophageal pitch matching

To be able to match pitch values, it is necessary to identify the treated phoneme. By analyzing pitch values of different esophageal phonemes, we notice that they have the same range of pitch values. We hence propose to add the two formant values F1 and F2 as criteria to identify the analyzed phoneme. As illustration, the table 24.1 resumes formant mean values for french vowels. These results are obtained with a long training database.

Table 24.1. Formant and pitch values for french vowels.

	a	o	u	i	y
F0(natural)	85	75	110	110	90
F0(esophageal)	50..140	50..120	50..140	50..140	50..140
F1	1000	500	320	320	320
F2	1400	800	3200	1650	1800

Once the phoneme identified, we replace the pitch already calculate by the value of the same phoneme pitch in the natural voice. This is what is called matching. By this way, the pitch values in the figure 24.3.b will be similar to the pitch outline of 24.3.a. These new values of the pitch will be used for the source synthesis.

24.3.3 Source synthesis

The natural voiced speech source signal is perfectly represented by the Liljencrants-Fants model[4] which is the most used. For this reason, we choose it to synthesis the esophageal source's signal. For the unvoiced frame, w use a white gaussian noise signal.

Moreover, as it known, the esophageal speech is characterized with a low intensity. So, we propose to amplify the excitation signal by multiplying its amplitude y a factor α. After intensive test, we choose $\alpha = 1.2$. Figure 24.4 shows the original (top figure) and the synthesized (bottom figure) excitation signal for the french sequence "aoa". Original source signal is caracterized y his large dynamics, without any noticed periodicity. However, synthesised source represent a curve caracterized by low dynamics and a regular form which is our main goal. Indead, we seek to construct a pariodic source signal with a natural pitch.

In term of perceptual listening quality, we notice an improvement of the esophageal speech mainly during vowels prononciation which become more distinguishable. Nevertheless, the perception of this improvement needs a lot of concentration during listening. We move now to the second part of the enhancement which deals with the vocal track enhancement.

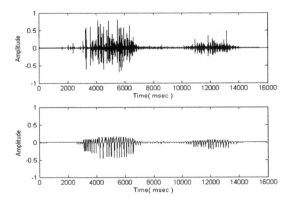

Fig. 24.4. Original and synthesized source signal.

24.4 Formantic structure enhancement

24.4.1 Bandwidth expansion

The vocal track signal is characterized by its formants and a formant analysis will guide us to vocal track enhancement. First, we extract the formant and their related bandwidth according to the technique illustrated on the flowchart 24.5.

Then, we compare the values of the natural formants and the esophageal formant. We observe that the frequency of the esophageal formant and the natural formant have the same range of values. However, when we focus on the bandwidth of formant, we remark that the natural bandwidth are larger than the esophageal bandwidths. The figure 24.6 shows the difference. Hence, a first idea of enhancement consist on enlarging the bandwidths. That's why, we propose the following approach. The vocal track are modelled by a cascade of second order resonator whose transmittance is described by:

$$W(z) = \frac{\sigma}{A(z)}; \cdot \tag{24.1}$$

where σ is the amplitude parameter, $A(z)$ is a plolynomial discribing the z transform of the vocal track and a_i are the linear prediction coefficients.

Each second order resonator reflects a formant. So then, the transmittance of the resonator is responsible for the formant values. The resonator bandwidth is:

$$\rho_k = \frac{(1 - p_k)f_e}{\Pi}, \tag{24.2}$$

where f_e is the frequency of sampling, p_k is the module or the $A(z)$'s roots. If we change the resonator and his transmittance became:

$$H'(z) = \sigma / A(z/\gamma), \tag{24.3}$$

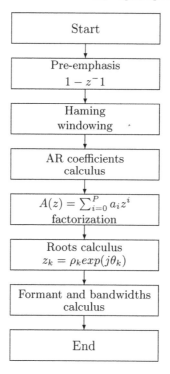

Fig. 24.5. Formant extraction flowchart

where σ is a constant to be adjusted.

The modification will affect the bandwidth. More precisely, it will be extended according to the following expression:

$$\rho'_k = \frac{(1-\gamma p_k)f_e}{\Pi} = \rho_k + p_k(1-\gamma)/\Pi; . \qquad (24.4)$$

This relation shows that the bandwidth passes from ρ_k to ρ'_k and the amount of expansion equals $\rho_k(1-\gamma)/\pi$ and depends on γ.

24.4.2 Reduction of spluttering noise

Thanks to formant bandwidth expansion, intensive listening test show that esophageal speech is well amplified and becames more intelligible. However, the background noise, when it appears is also amplified which degrades slightly the listening quality. So, a second idea of enhancement takes form. We have to amplify the low frequencies where useful speech is present and leave constant

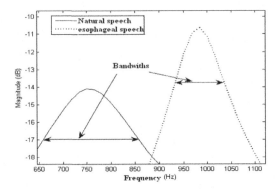

Fig. 24.6. Bandwith of natural voice and esophageal voice .

high frequencies where the noise is predominant. We propose a correction for
the used perceptual filter. In fact, we can use an another filter defined with:

$$H'(f) = \begin{cases} H(f) \text{ if } f \leq f_{change} \\ 1 \text{ otherwise,} \end{cases} \qquad (24.5)$$

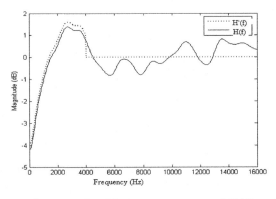

Fig. 24.7. An example of frequency response of $H(f)$ et $H'(f)$.

with f_{change} is the limit of the low frequencies. The figure 24.7 illustrate
an example of the frequency response of the first and the second filter. The
figure 24.8 shows the

24.5 Results and discussion

These treatment can be englobed in three different manner:

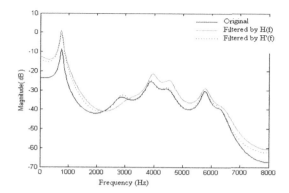

Fig. 24.8. Signals before and after the filtering.

1. Schema I: We treat the signal in the same chronological way explained in this paper: we enhance the source signal and then wee enhance the formantic structure of the signal.
2. Schema II: in this way, the enhancement of the esophageal speech begins with the amelioration of the formantic structure then the enhancement of the glottal source signal.
3. Schema III: these enhancements are made at the same time.

Subjective evaluation test have been made by 15 persons (7 persons well used to listening the esophageal speech and 8 persons who have never hear it before) who listened to the sequences of our data base before enhancement. The sequences used for the evaluation are french sentences:

1. aoa.
2. C'est un poison pour les poissons.
3. Charitbien ordonn commence par soi meme.
4. Ciel si ceci se sait, ces soins sont sans succes.
5. Prenez vos papiers de votreporte-feuille.
6. Tout le monde y songe.

The table 24.2 sums up the listening test result. For every sequence and every method, we found the rate of appreciation. It's quite clear that the enhanced sentences are more appreciated than the original signal. We can notice that the second method is the most appreciated with a rate of 66%. The first or the third method could also be more liked depending on the sentence. Some times, the three method gives sentences quite similar. That's why, in 8% of listeners can't defferenciate the 3 enhancement.

Table 24.2. Listening quality evalution of proposed enhancement techniques.

Sequences	original signal	MethodI	MethodII	MethodIII	No difference
1	0 (0%)	3 (25%)	5 (42%)	3 (25%)	1 (8%)
2	0 (0%)	5 (42%)	4 (33%)	2 (17%)	1 (8%)
3	1 (8%)	3(25%)	3 (25%)	3 (25%)	2 (17%)
4	1 (8%)	2 (16.33%)	5 (42%)	2 (16.33%)	2 (16.33%)
5	1 (8%)	4 (33%)	5 (42%)	3 (25%)	1 (8%)
6	0 (0%)	3 (25%)	5 (42%)	3 (25%)	1(8%)

24.6 Conclusion

In this paper, a esophageal enhancemnet unsing source synthesis and formant pattern is presented. First, we presented, the pitch extraction with the EV-MACM. Then, we correct the pitch values to the natural pitch values. THese modified pitch values are used for the source signal synthesis. In the second part, the formant bandwith are enlarged. These enhancement are then englobed in three different ways.

References

1. Gualberto AGUILAR, Mariko Nakano-Miyatake, Hector Perez-Meana: Ala-ryngeal Speech Enhancement Using Pattern Recognition Techniques. IEICE TRANS. INF. &SYST. (2005)E88-D
2. A Hisada, H Sawada: Real-time clarification of esophageal speech using a comb filter. International Conference Series On Disability, Virtual Reality And Associated Technologies,
3. A Hisada, N Takeuchi, H Sawada: Real-time clarification filter of a dysphonic speech and its evaluation by listening experiments. Intl Conf. Disability. Virtual Reality & Assoc. Tech.(2004)
4. Gunnar Fant, Johan Liljencrants, Qi-guang Lin: A four parameter model of glottal flow. French-Sweden symposium,Grenoble,

Arbitrary Image Cloning

Xinyuan Fu, He Guo, Yuxin Wang, Tianyang Liu, Han Li

Department of Computer Science and Technology, Dalian University of Technology, Dalian, China, 116024 `guohe@dlut.edu.cn/peteryp@126.com`

Summary. In this paper, an image cloning method called arbitrary cloning is presented, in which image matting and Poisson image editing techniques are applied. The object boundary is extracted through image matting technique while Poisson equation is solved to achieve natural result for seamless cloning. Compared with existing techniques, the proposed approach gains a great deal of advantages: On the one hand, great performance can be reached even in bad conditions. The method will not be affected even when there are many holes in the foreground image and there is complicated color variation in the foreground and background image. On the other hand, the algorithm is very flexible in application that the α matte can be guided by the user input. The examples show that the proposed approach guarantees better output and diverse result.

Key words: image matting; Poisson equation; gradient field; arbitrary cloning

25.1 Introduction

Image seamless cloning is to separate the target object from source image and then embed the selected object into the target image. Such a technique is placed in a core position in image editing field. After Patrick Perez and Michel Gangnet et al.[1] developed an approach of image cloning, there has been continuing interest in the development of novel types of image cloning algorithm based on both edge detection and color cloning technique.

Poisson editing method makes seamless editing of image regions by solving Poisson equations with Dirichlet boundary conditions. The Poisson equation can be discretized by Laplacian 5-point finite difference formula, and the optimal solution is computed by Gauss-Seidel iteration. Different initial value, iterations and guidance gradient field can lead to different cloning result. Although the Poisson editing algorithm is a compact algorithm and easy to understand, it must detect the edge of target object manually, which is its biggest disadvantage. When the shape of the target in the source image is complicated, detecting the image edge manually is very difficult and a little error may lead to bad cloning result (Fig.25.1(e)(f)). Therefore, it is important to extract the target image edge accurately in image cloning method. In this paper, an image seamless cloning method which combines the image matting and Possion editing techniques is proposed. Such an approach can achieve more rapid cloning speed and better result compare to existing cloning algorithms.

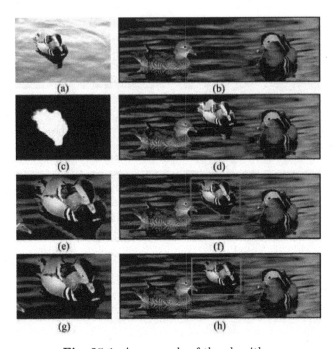

Fig. 25.1. An example of the algorithm

Formally, image matting method satisfies the following model:

$$I_i = \alpha_i F_i + (1 - \alpha_i) B_i \qquad (25.1)$$

where α_i is the pixel's foreground opacity, F_i is the foreground image and B_i is the background image.

To extract α, F and B, for a RGB color image, User constraints are added in to extract a good matte. Current methods need the user to input a trimap. These methods typically use non-linear iteration to obtain α, F and B. This will cost much time for a large unknown region in a trimap. If the image is complex in structure, user must spend more time and energy to provide a good trimap. L. Anat and L. Dani et al.[2] presented a closed-form solution for extracting the alpha matte from a natural image. It optimized a cost function from local smoothness assumptions on foreground and background colors, and eliminate F and B yielding a cost function in α. The α matte can be obtained by solving a sparse linear system. A small amount of user input is needed for extracting a high quality matte. Due to the flexibility of user input, the foreground matte varies with different user input.

In this paper, a new image cloning algorithm is proposed which needs only α as the gradient field mark of Poisson equation. This algorithm will not compute F and B and therefore will decrease the computing cost. The new image cloning solution combines two different types of algorithm which can achieve a better cloning result.

Fig.25.1 shows a cloning result using the proposed technique on a sample image and compares this result to Poisson Image Editing algorithm. (a) is a source image and (b) is a target image. (c) is a alpha matte exacting from source image. (d) is a cloning result which combines two images directly. (f) is a Poisson Image Editing result and (e) is its partial enlargement. (h) is a cloning result with proposed method and (g) is its partial enlargement with natural boundary and seamless cloning.

This paper is organized as follows. Section 2 discusses the related work. Section 3 introduces foreground matte extraction. Section 4 introduces Poisson image editing algorithm. Section 5 presents the proposed cloning solution and discusses several implementation issues. Section 6 evaluates the experimental results and compares with other algorithms. Finally in section 7, concluding remarks are offered.

25.2 Related Research

Image matting and image cloning are two different fields in image processing. Image matting is to extract α, F and B from a given image. The tone blend is not very important in image matting but very important in image cloning. Image cloning does not consider the quality of exacting foreground object but image matting does.

J. Jiaya and S. Jian et al. presented a solution for seamless image composition [3]. They used shortest closed-path algorithm to search for the location of the boundary; then they used a blended guidance field for Poisson equations to extract the fractional boundary of the object. This approach achieves seamless image cloning with only small amount of user input. When the source and target images are similar in color, it may not work well. Furthermore, if the target image has complex structures, the structure of the source region and target scene cannot be precisely aligned. The user cannot control the foreground matte directly. If there are many holes in the image, it may be slow because the algorithm repeats the same work for several times. It also takes extra time to compute α and α mask (M).

Recently, some new algorithms for extracting a foreground region from background have been successfully proposed. They commonly solve α, F and B with additional constraints, others with flashlight image and active-light image. J. Wang and M. Cohen proposed an iterative optimization approach to solve the image matting problem [4]. This approach uses an iterative non-linear optimization algorithm. J. Sun and J. Jia et al. proposed Poisson matting for image matting [5]. In their approach, the matte is directly reconstructed from a continuous matte gradient field by solving Poisson equations using boundary information from a user-supplied trimap. It involves local Poisson matting which requires many user interactions. [6] is regarded as a classical method. It transforms the matting problem to a Bayesian framework, and extracts α matte. Other approaches [7][8] also achieve good result. These approaches require constructing trimap as input, which is a difficult work itself. Some algorithms have been presented to help user construct trimap [9][10]. They obtain a "hard" segmentation using iterative graph cut, and a coarse trimap can be gained from the result. These algorithms are not stable and may lead to wrong result if the image is complex enough.

As far as we know, there are three existing seamless cloning techniques. The first one is used in Adobe Photoshop, which has not been published. The second one is the multi resolution image blending proposed in [11]. It builds Laplacian pyramid and interpolates to interfuse two images. The third one is briefly discussed in Section 1 and will be detailed in Section 4, which is closely related to the proposed approach.

25.3 Matte Extraction

25.3.1 Basic Theory

First, the matte is extracted through image matting technique. The problem is under-constrained, so some basic assumptions are made on α, F and B[2].

For grayscale image, it assumes that in the neighborhood window of each pixel, F and B are nearly constant. So (25.1) is rewritten as:

$$\alpha_i \approx aI_i + b, \forall i \in \omega \qquad (25.2)$$

Where, $a = \frac{1}{F-B}, b = -\frac{B}{F-B}$, ω is a set of windows. The algorithm finds α,a,b to minimize the following function[2]:

$$J(\alpha, \mathbf{a}, \mathbf{b}) = \sum_{j \in I}(\sum_{i \in \omega_j}(a_i - a_j I_i - b_j)^2 + \epsilon a_j^2) \qquad (25.3)$$

Where, ω is a small window around j. It is a quadratic function with respect to α,a,b. Since, it's hard to solve the optimization problem directly, according to [2], a,b can be eliminated, which results in a function with only one parameter α: $J(\alpha) = min_{\mathbf{a},\mathbf{b}}J(\alpha, \mathbf{a}, \mathbf{b}) = \alpha^T L \alpha$, α is $N \times 1$ vector, L is a $N \times N$ matrix, with N pixels. The (i, j)-th member of L matrix is:

$$\sum_{k|(i,j)\in\omega_k} (\delta_{ij} - \frac{1}{|\omega_k|}(1 + \frac{1}{\frac{\epsilon}{|\omega_k|} + \sigma_k^2}(I_i - \mu_k)(I_j - \mu_k))) \qquad (25.4)$$

Where, δ_{ij} is kronecker delta. μ, σ^2 is the mean and the variance of the pixels in the window ω_k around k. $|\omega_k|$ is the number of pixels in the window.

For color image, [2] relaxes the basic assumption, which gains advantage that the new color model satisfies most of images, so it can be applied in more conditions. And it finally derives the formula $J(\alpha) = \alpha^T L \alpha$, with L an $N \times N$ matrix. The (i,j)-th member of L is:

$$\sum_{k|(i,j)\in\omega_k} (\delta_{ij} - \frac{1}{|\omega_k|}(1 + (I_i - \mu_k)(\Sigma_k + \frac{\epsilon}{|\omega_k|}I_3)^{-1}(I_j - \mu_k))) \qquad (25.5)$$

where, Σ_k is a 3×3 covariance matrix, μ_k is a 3×1 mean vector in a window ω_k and I_3 is a 3×3 identity matrix. The contribution of ϵ and $|\omega_k|$, as well as the comparison between L and classical affinity function $W_G(i,j) = e^{-||I_i - I_j||^2/\sigma^2}$, is analyzed completely in [2]. Because the eigenvectors of L are used in image segmentation[12], the smallest eigenvectors of L can guide the user where to place the input information.

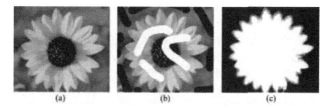

Fig. 25.2. Extraction of α matte

25.3.2 Add User Constraints

First, the algorithm needs the user to mark the foreground and background roughly. It's scribbled in white and black distinguishingly, as shown in Fig.25.2(a)(b). Secondly, the constraints are added into the optimization problem:

$$C(\alpha) = \alpha^T L \alpha + \lambda(\alpha^T - b_s^T)D_s(\alpha - b_s) \qquad (25.6)$$

where, the second term is the user constraints, λ is a large number that decides the contribution of the user constraints in the whole function, D_s and b_s are detailed in [2]. All the partial derivatives of the function are set to 0 to gain a linear system of equations (25.7). So α matte is extracted by solving the equation, as shown in Fig.25.2(c).

$$(L + \lambda D_s)\alpha = \lambda b_s \qquad (25.7)$$

25.4 Cloning Theory

25.4.1 Basic Theory

We derive in the single color channel and extend to the RGB color space. The goal is to interfuse the region Ω in source image I_s into target image I_t, f is an unknown function defined in $\Omega - \partial\Omega$. $\mathbf{v}$ is a known vector defined in the field Ω in I_s. [1] shows an approach to solving the following optimization problem for interfusing two images:

$$min_f \int\int_\Omega |\nabla f - \mathbf{v}|^2 \ with \ f|_{\partial\Omega} = I_t|_{\partial\Omega} \tag{25.8}$$

$\nabla. = [\frac{\partial.}{\partial x}, \frac{\partial.}{\partial y}]$ is a gradient operator. According to the Ostrogradskii equations in calculus of variations theory[13], the solution is Poisson equation with Dirichlet boundary conditions

$$\Delta f = div \ \mathbf{v} \ over \ \Omega, \ with \ f|_{\partial\Omega} = I_t|_{\partial\Omega} \tag{25.9}$$

$div \ \mathbf{v} = \frac{\partial u}{\partial x} + \frac{\partial v}{\partial y}$ is the divergence of $\mathbf{v} = (u, v)$, $\Delta. = \frac{\partial^2.}{\partial x^2} + \frac{\partial^2.}{\partial y^2}$ is the Laplacian operator.

25.4.2 Discrete Presentation

The discretization problem can be solved using image pixels. For each pixel p, N_p is a set of its 4-connected neighbors, $< p, q >$ denotes a pixel pair, with $q \in N_p$. f_p is f value on point p, and the target is to find $f|_\Omega = \{f_p, \ p \in \Omega\}$.

First, the gradient field needs to be defined, let $\mathbf{v} = \nabla I_s|_\Omega$, with $\mathbf{v}_{pq} = I_{sp} - I_{sq}$ as the discretized presentation. So (25.9) is now:

$$\Delta f = \Delta I_s \ over \ \Omega, \ with \ f|_{\partial\Omega} = I_t|_{\partial\Omega} \tag{25.10}$$

According to Laplacian 5-point finite difference formula, (25.10) is discretized as:

$$|N_p| - \sum_{q\in N_p\cap\Omega} f_q = \sum_{q\in N_p\cap\partial\Omega} I_{tq} + \sum_{q\in N_p} \mathbf{v}_{pq}, \ for \ all \ p \in \Omega \tag{25.11}$$

When Ω contains the boundary pixels, the 4-connected neighbor becomes 3 or 2-connected, so $|N_p| < 4$. But for the pixels inside Ω, (25.11) is rewritten with no boundary conditions $|N_p| - \sum_{q\in N_p} f_q = \sum_{q\in N_p} \mathbf{v}_{pq}$. Due to the randomicity of the boundary conditions, it's hard to create a linear system of equations. So linear iterating such as Gauss-Seidel is used to optimize the problem.

25.5 Arbitrary Image Cloning

The work of this paper is to combine the two kinds of algorithms introduced above. An interactive system is designed to apply the seamless cloning technique. The user first scribbles the foreground and background image roughly using any of the painting software, and then extract the matte. After that the user chooses where to place the source region into the target image so that the interfusing operation is done.

The matte extraction needs to set up the matrix L, and solve a linear system. The matte is equal to a grayscale image whose value is between 0 and 1, and it represents the transparency of the foreground image. We can use it as a mark table, and only the corresponding pixels in the source image will be interfused into the target image.

There are several methods to define the gradient field. [3],[1] gave two different methods. The former takes the blend of source and target image with α information as gradient field and the latter compares the absolute value of the gradient between the corresponding pixels in source and target images, takes the larger as the gradient of the pixel.

In this paper, the Ω region of the source image is chosen directly as the gradient field for image seamless cloning (The results are satisfied as the experiments shown in Section 6).

The region of Ω is defined as:

$$x \in \Omega, \; if \; \alpha(x) > AlphaThresh \qquad (25.12)$$

where $AlphaThresh$ is a user defined parameter which is set to 0.3 in this paper.

The gradient field of seamless cloning is:

$$\mathbf{v}(x) = \nabla I_s(x), \; x \in \Omega \qquad (25.13)$$

For image blending, the source image is not totally interfused into the target image. And the effect is to superpose the colors of two images. The classical method is to compute the weighted summation of RGB value of two images, which leads to bad result and unnatural boundary. To solve this problem, the following gradient field is taken:

$$\mathbf{v}(x) = \begin{cases} \nabla I_s(x), \; if \; |\nabla I_s(x)| > |\nabla I_t(x)|, \; x \in \Omega \\ \nabla I_t(x), \; otherwise, \; x \in \Omega \end{cases} \qquad (25.14)$$

Fig.25.3 shows the difference between the two methods. (a)(b)(c)(d) are distinguishingly the source image, the use input, the α matte and the target image. (e)(f) show the results using the first kind of gradient field, while (g)(h) using the second kind of gradient field. The difference between (f) and (h) can be observed. (f) takes the source image as the gradient field, so the result is not blended with the target image. (h) takes the blended gradient field so it contains the color variation of both images.

With the gradient field, Gauss-Seidel iteration is applied. The initial value can be the source image or the target image. Based on different application, this will gain different results. In addition, the user can control the number of iterating, this also generates a variety of results. The next section will give an analysis in detail.

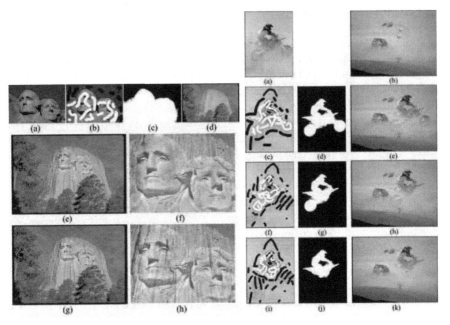

Fig. 25.3. Image cloning and blend

Fig. 25.4. The role of user guided information

25.6 Experiments

[3] has got excellent results, in comparison, the arbitrary cloning algorithm proposed in this paper gains advantage that the shape of the foreground matte can be roughly controlled by users. Due to patent problem, we cannot get the executable file of the paper [3]. So the comparisons are made with images from paper [3] and the author's web site directly. Comparisons are also made with Poisson Image Editing and direct compositing. Experiment results indicate that the proposed algorithm gain better result. Several experiments were carried out on different purpose. And they can verify the flexibility and accuracy of the algorithm.

Experiment 1 demonstrates that the foreground matte can be guided by user to produce flexible cloning effects. And this behavior makes the primary advantage of the algorithm proposed in this paper. In Fig.25.4, (a)(b) are the source and target image. (c)(f)(i) are user inputs, and (d)(g)(j) are the foreground matte according to (c)(f)(i). It is obvious that the foreground matte varies with different user input. In (d), the driver and the whole motorcycle are extracted. In (g), the front wheel of the motorcycle is considered as background. Only the driver, motorcycle's body and the back wheel are extracted. In (j), only the person and motorcycle's body are extracted. Different foreground are interfused into the same target image, and the final results are shown in (e)(h)(k). The user input can be changed according to different user demands and applications. No matter how the user input changes, the algorithm achieves excellent result. All the results in Fig.25.4 use the target image as the initial value and iterate for 200 times.

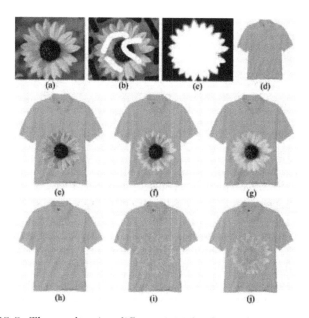

Fig. 25.5. The result using different initial value and iterating number

Experiment 2 shows different cloning results using different initial value and iterating number, as shown in Fig.25.5. (a)(b)(c)(d) are a source image, a user input, a foreground matte and a target image. To avoid the influences of inconsistent tone, the source image is transformed into grayscale image. (e)(f)(g) use the source images as initial values, and the iterating number is 10, 200 and 2000 times. (h)(i)(j) use the target images as initial values, and the iterating number is 10, 50 and 200 times. If the iterating number is large enough, the final results are the same using either initial value. While using the source image as initial value, the cloning tone is much similar to the source image, and vice versa.

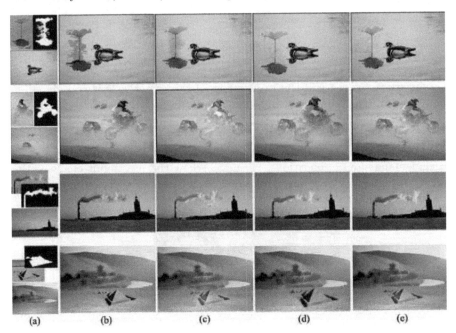

(a)   (b)   (c)   (d)   (e)

Fig. 25.6. Comparison between different algorithms

Comparisons between different algorithms are shown in Experiment 3. Fig.25.6 uses the images in [3] as source images and target images, and all the comparisons between Poisson Edit, direct compositing and our paper are based on these images. Column (a) shows the source images, foreground mattes and target images. The results of direct compositing are shown in column (b). Poisson Edit results are indicated in column (c), where the cloning results are unnatural due to the uncertainty of the boundary conditions. Results of algorithm [3] are shown in column (d) and the proposed method in column (e). The results indicate that both (d) and (e) produce satisfying results, using the images in [3] as testing images. Experiment 1 proves that our algorithm is more flexible and powerful in bad conditions, such as complicated color or boundary variation. The arbitrary image cloning algorithm can be used in various situations, while other popular algorithms are restricted to a certain kind of images.

Experiment 4 achieves image blending using the blended gradient discussed above. Besides, some interesting results are obtained. And all these results show that the algorithm can be used in various applications, and can also provide better image seamless cloning solutions for users. The image cloning algorithm can be used in the fields of art and innovation design, Fig.25.7 shows an interesting result. The water lily can be blended with (a) or (b). However, the Da Vinci self-portrait can only be blended with (b). That's because for most of the pixels in (a), the absolute value of the gradient is larger. When

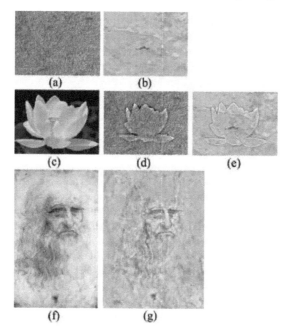

Fig. 25.7. Image blend

(a) and (f) is used, the blended gradient field is equal to (a), thus there is no obvious effect. The gradient of (b) is much smoother, so most of the images can be blended with (b).

The sparse linear system of equations $(L + \lambda D_s)\alpha = \lambda \mathbf{b}_s$ is solved by LU method, where is a symmetrical semi-definite matrix. It will take several seconds when calculating the foreground matte of a 400×300 image on P4 2.8GMhz, 768MB memory computer. And during the Gauss-Seidel iteration, the calculating time depends on the iterating number and the error threshold. Generally, for a 200×200 iterating region, it takes about 4 seconds to iterate for each 100 times.

25.7 Conclusion

The paper presents an arbitrary cloning method. Image matting and Poisson image editing techniques are applied to accomplish the image seamless cloning and to achieve a complete solution for image interfusing. The object boundary is extracted through image matting technique while Poisson equation is solved to achieve natural result for seamless cloning. A number of experiments indicate that the paper gains advantages in performance under bad conditions.

The main work of this paper is that two different kinds of algorithms are combined to solve this challenging problem, and a framework of combination is proposed. But it's not satisfying when dealing with large dataset. The cost-ing time of foreground extracting increases saliently with the size of image. The reason is that the L matrix is so large. So if the user chooses a image of 800×600 or bigger on common memory condition, the matting problem transforms into solving the sparse linear system of equations, which was a well studied subject. Our future research directions are to improve the speed of the algorithm and apply a similar technique on video seamless cloning.

References

1. Perez, P., Gangnet, M., Blake. A.: Poisson image editing. Proceedings of ACM SIGGRAPH. New York, USA, ACM Press, (2003) 313-318
2. Anat Levin, Dani Lischinski, Yair Weiss.: A Closed Form Solution to Natu-ral Image Matting. IEEE Computer Society Conference on Computer Vision and Pattern Recognition. New York, USA, IEEE Computer Society, (2006) Volume1, 61-68
3. Jiaya Jia, Jian Sun, Chi-Keung Tang, Heung-Yeung Shum.: Drag-and-drop pasting. Proceedings of ACM SIGGRAPH. New York, USA, ACM Press, (2006) 631-637
4. J. Wang, M. Cohen.: An iterative optimization approach for unified image seg-mentation and matting. Proc. IEEE Intl. Conf. on Computer Vision. Beijing, China, IEEE Computer Society, (2005) Volume2, 936-943
5. Sun, J., Jia, J., Tang, C., Shum, H.: Poisson matting. Proceedings of ACM SIGGRAPH. New York, USA, ACM Press, (2004) 315-321
6. Chuang, Y., Curless, B., Salesin, D., Szeliski, R.: A bayesian approach to digital matting. In: Proceedings of CVPR. New York, USA, ACM Press, (2001) vol. 2, 264-271
7. Ruzon, M., Tomasi, C.: alpha estimation in natural images. Proceedings of CVPR. New York, USA, ACM Press, (2000) 18-25
8. Berman, A., Vlahos, P., Dadourian, A.: Comprehensive method for removing from an image the background surrounding a selected object. U.S. Patent: 6,134,345, (2000)
9. [9] Li, Y., Sun, J., Tang, C.,Shum, H.: Lazy snapping. In: Proceedings of ACM SIGGRAPH. New York, USA, ACM Press, (2004) 303-308
10. Rother, C., Kolmogorov, V., Blake, A.: "grabcut" - interactive foreground ex-traction using iterated graph cuts. Proceedings of ACM SIGGRAPH. New York, USA, ACM Press, (2004) 309-314
11. Burt, P., Adelson, E.: A Multiresolution Spline with Application to Image Mo-saics. ACM Transactions on Graphics. New York, USA, ACM Press, (1983) 2, 4, 217-236
12. J. Shi, J. Malik.: Normalized cuts and image segmentation. Proc. CVPR. San Juan, Puerto Rico, IEEE Computer Society, (1997) 731-737
13. Wu Diguang.: Calculus of Variations. Beijing: Higher Education Press. (1987) 65-80
14. Wang Wenxian, Zhang Fengxiang, Shi Guangyan.: Linear Algebra and Its Ap-plication. Dalian University of Technology Press, (1993) 40-59

Iterative Joint Source-Channel Decoding with source statistics estimation: Application to image transmission

Haifa Belhadj[1], Sonia Zaibi[2] and Ammar Bouallègue[3]

[1] SYSCOM Lab, ENIT (Tunisia),haifa.belhadj@yahoo.fr
[2] SYSCOM Lab, ENIT (Tunisia),sonia.zaibi@enit.rnu.tn
[3] SYSCOM Lab, ENIT (Tunisia),ammar.bouallegue@enit.rnu.tn

Summary. This chapter deals with iterative joint source channel decoding and its application to image transmission. First, the problem of transmitting a correlated gaussian source over an AWGN channel is considered. The joint decoding is implemented by the Baum Welch algorithm estimating the source statistics. Iterations between the MAP channel decoder and the source decoder are made to improve the global decoder performance. This decoding scheme is then applied to an image transmission system, based on a wavelet decomposition of the source image followed by a DPCM coding of the lowest frequency subband and a SPIHT coding of high frequency subbands. Simulation results show that a significant performance gain is obtained with iterative joint source channel decoding, compared to a classical decoding, in case of a correlated gaussian source and also in case of image transmission.

Key words: Joint Source-Channel Decoding, Iterative Decoding, Baum-Welch Algorithm, DWT, DPCM, SPIHT.

26.1 Introduction

In traditional communications systems, source and channel coding are performed separately. However, the separation between source and channel coding has turned out to be not justified in practical systems due to limited coding/decoding delay and system complexity. On these circumstances, one can improve performance by considering the source and channel design jointly. Research on this area goes back to the work of Fine [1] and continuous to the present [2]. On the other hand, Turbo-codes [3], with their iterative decoding techniques, achieve very good performance, which are close to the theoretical limit of Shannon. In this paper, we consider the problem of transmitting a correlated source over an AWGN (Additive White Gaussian Noise) channel. This source is protected by a convolutional code. We use the turbo codes principle to release an iterative joint source channel decoding algorithm with estimation of source statistics. This scheme is applied to an image transmission system. In chap26section1, we briefly remind the Baum-Welch algorithm principle, applied to joint source-channel decoding. The turbo decoding principle is used in the chap26section2, to release an Iterative Joint Source Channel Decoding (IJSCD) algorithm. Performances of iterative decoding for a correlated gaussian source transmission are presented in chap26section3. In chap26section4, the IJSCD is applied to an image transmission system, based on a wavelet decomposition of the source image followed by a DPCM (Differential Pulse Code Modulation) coding of the lowest frequency subband and a SPIHT (Set Partitioning in Hierarchical Trees) coding for high frequency subbands. Finally, simulation results for the lowest frequency subband and the entire image, are respectively given. Chap26section5 draws conclusions and suggests future work.

26.2 Joint source-channel decoding and estimation of correlated source parameters

We consider the problem of encoding and transmitting a source signal vector $I = \{i_0, i_1, , i_t, , i_{T-1}\}$ over a noisy channel. We still want to know the sequence of transmitted source indexes it but they are not directly observable because of the possible corruption by the channel. Instead, we have the received indexes, $O = \{o_0, o_1, ., o_t, .o_{T-1}\}$, which are the observations related to the input probabilistically. This situation can be directly interpreted as a discrete Hidden Markov Model (HMM). A discrete HMM can be defined by two parameters and three probability matrices. The parameters are K the number of states, and T the source sequence length[4]. To determine, at each time, the most likelihood symbol, we use the BCJR (Bahl Cocke Jelinek and Raviv) algorithm originally proposed in [5] and based on the forward-backward algorithms. The BCJR allows to calculate the a posteriori probability denoted $\gamma_t(i)$:

$$\gamma_t(i) = P[I_t = i | O, \lambda] \tag{26.1}$$

To calculate $\gamma_t(i)$, we need to determine two variables: $\alpha_t(i)$ and $\beta_t(i)$. The BCJR algorithm combines the forward induction with the backward one to compute the probabilities $\gamma_t(i)$. For more details refer to [5]. The methods above allow to determine the most likelihood a posteriori symbol, where $\gamma_t(i)$ is maximum, with consideration of a hidden Markov source whose parameters are known. A more powerful approach would allow the receiver to use the noise-corrupted observations available in the decoder to estimate the parameters characterizing the hidden Markov source. We will estimate source statistics by the Baum-Welch algorithm called also EM [6][7].

Estimation of source parameters at the receiver

We try to estimate the source transition matrix A. Its elements $a_{i,j}$ are the source transition probabilities:

$$a_{i,j} = P(I_t = j|I_{t-1} = i); 0 \le i, j \le K - 1 \tag{26.2}$$

To do that, we introduce a new parameter $\psi_t(i, j)$ representing the probability that source state is i at the time t and j at time $t+1$:

$$\psi_t(i,j) = P[I_t = i, I_{t+1} = j|O, \lambda] = \frac{\alpha_t(i)a_{i,j}P(O_{t+1} = o_{t+1}|I_{t+1} = j)\beta_{t+1}(j)}{\sum_{i=0}^{K-1} \alpha_t(i)\beta_t(i)} \tag{26.3}$$

The re-estimation formula is then given by:

$$a_{i,j} = \frac{\sum_{t=0}^{T-2} \psi_t(i,j)}{\sum_{t=0}^{T-2} \gamma_t(i)} \tag{26.4}$$

After having re-estimated the parameter A of the initial hidden Markov model, the algorithm will repeat iteratively the re-estimation with the new model (we calculate another time the α and β values).This process can be repeated iteratively until no further improvement in the model results. The transition source matrix A is initialised as follows: $a_{i,j}^0 = P[I_t = j]$. After calculating the *a posteriori* probabilities, we determine at each time t, the value $\widehat{i}_t$ maximising $\gamma_t(i)$. $\widehat{i}_t$ is the most likelihood symbol value at the time t which will be decoded.

26.3 Iterative joint source-channel decoding

26.3.1 System model

We consider the system model shown in chap26figure1. The correlated source produces a sequence of T continuous-valued, gaussian distributed symbols, with a variance equal to 1 and a correlation factor equal to 0.9. Each symbol of the sequence is quantized by a scalar quantizer, that produces a sequence of indices $I = (i_0, ..., i_t, ...i_{T-1})$ According to a fixed length bit mapping, each index i_t is assigned a unique binary sequence $B_t = (b_{t,1}, ..., b_{t,L})$which generates a bit sequence $B = (B_0, ..., B_t, ...B_{T-1})$ of length $K = T * L$ bits, where L is the binary code word length. This bit sequence is bitwise interleaved with an interleave denoted Π, before being coded by a recursive systematic convolutional encoder and transmitted over an AWGN channel using BPSK modulation. The received sequence is denoted $Y = (y_0, ..., y_t, ..., y_{T-1})$; $y_t = (y_{t,1}, ..., y_{t,l}, ..., y_{t,L})$ It constitutes the iterative decoder input.

Both source and channel decoders are Soft In/Soft Out (SISO). The MAP algorithm is used as a channel decoder, while the Baum-Welch (EM) algorithm is used as a source decoder estimating source statistics. The Iterative Joint Source Channel Decoding (IJSCD) method will now be described.

26.3.2 Iterative Joint Source Channel Decoding (IJSCD) method

The iterative decoding scheme is related to turbo decoding. It consists of a data exchange between two or more channel decoders which are SISO decoders. The source and channel decoders receive and send their messages in

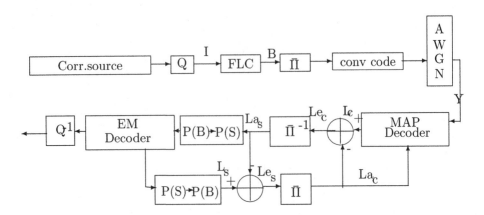

Fig. 26.1. System model with IJSCD (Q: Scalar Quantizer, FLC: Fixed Length Bit Mapper, Π : Interleaver)

terms of Log-Likelihood Ratio (LLR). Let's remind that we have an information exchange between the source decoder, operating with symbol data, and the channel decoder working with bit data. So we need conversion blocks P(S)→P(B) and P(B)→P(S), which allow to calculate bit probabilities from symbol ones and back again. The channel decoder allows us to determine an a priori information Las for the source decoder. However, this information is a bit information and the Baum-Welch algorithm needs a symbol one. The conversion bloc P(B)→P(S) is used to calculate symbol probabilities from bit ones. Formally, if we write the probabilities for each bit input to the P(B)→P(S) block as , where $c \in \{0,1\}$, then symbol probabilities are approximated by the product of the corresponding bit probabilities:

$$P(I_t = i) = \prod_{l=1}^{L} P_A(b_{t,l} = map_l(i_t)) \qquad (26.5)$$

where $map_l(i_t)$ is the bit of position l in the bit word mapping the symbol i. The probability P_A is determined from the a priori information La_s:

$$La_s(b_{t,l}) = \ln(\frac{P_A(b_{t,l} = 1)}{P_A(b_{t,l} = 0)}) \qquad (26.6)$$

At the output of the source decoder, we need to know the bit probabilities to calculate the extrinsic information rescued at the channel decoder. The Baum-Welch algorithm used for the source decoding provides the a posteriori symbol probabilities. Therefore, we need to apply the P(S)→P(B) conversion. The bit probabilities are derived from the symbol ones as follows:

$$P(b_{t,l} = c) = \sum_{i/map_l(i)=c} P(I_t = i) \qquad (26.7)$$

Then,the source decoder output is given by:

$$L_s(b_{t,l}) = \ln(\frac{P(b_{t,l} = 1)}{P(b_{t,l} = 0)}) \qquad (26.8)$$

We subtract the *a priori* information values La_s from this information to get the extrinsic values Le_s, rescued, after interleaving, to the channel decoder as an *a priori* information La_c. This decoding procedure is repeated iteratively. We pull up when performances stop to improve.

26.4 Simulation results (Case of a correlated Gaussian source)

To evaluate the proposed system (chap26figure1) performances, we plotted the Bit Error Ratio (BER) evolution, as a function of the signal to noise ratio $Eb/N0$. The results achieved are compared to a transmission chain using a classical decoding scheme (without iterative joint source channel decoding), considered as a reference chain. This chain uses in fact, the Viterbi algorithm as a channel decoder. We have considered the transmission of a sequence of 400 symbols, issued from a one-order Markov Gaussian source, with a variance equal to 1 and a correlation factor equal to 0.9. Each symbol value is quantized by a one step uniform scalar quantizer. The quantized indexes belong to a source alphabet of size 7. The fixed length bit mapper (FLC) associates to each quantized index a 3-bit binary code word. We have used a recursive systematic convolutional code [8] with generator polynomials $(37, 21)$ and rate $\frac{1}{2}$. The used interleaver is a random one of size $20 * 60$. Simulation results are represented on chap26figure2:

- IJSCD: iter4 refers to a transmission chain with iterative joint source-channel decoding, with source perfect knowledge, at the forth iteration.
- IJSCD+EM: iter$i$ refers to a transmission chain with iterative joint source-channel decoding, with source statistics estimation, at the iteration i.

We notice that for a BER of 3.10^{-4}, a gain of 1.5dB in $(\frac{E_b}{N_0})$, is achieved by the iterative decoding system with source statistics estimation (DICSC+EM) comparing to the reference chain. The gain is more than 2dB, for the iterative decoding with perfect knowledge of source statistics. For a signal to noise ratio of 4dB, the BER is near 10^{-5} for the transmission system with iterative joint source-channel decoding (perfect knowledge statistics), and only about 3.10^{-4} for the reference chain. So we can conclude that the proposed transmission system, bring a strong gain in performances comparing to a classical system based on a separated source and channel decoding.

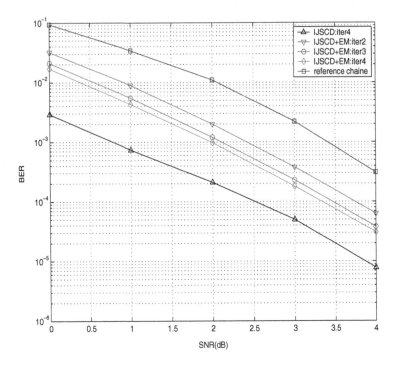

Fig. 26.2. BER as a function of $Eb/N0$ for transmission systems without and with (IJSCD) (with and without estimation)

26.5 Application to image transmission

The majority of efficient image compression algorithms use a transformation, applied to the original signal, a quantization and an entropy coding. According to the choice of the transformation, the quantization and the entropy coding, many compression schemes have been proposed. One of the most used transformations for image coding is the Discrete Wavelet Transformation (DWT)[9].

26.5.1 DWT and SPIHT principle

The discrete wavelet transformation is derived from the multiresolution analysis, developed by Stephane Mallat and Yves Meyer[10]. The aim of this theory is to decompose a signal into different resolutions. The lowest frequency subband contains the most important information of the image. The high frequency subbands constitute the image details. We use in our work, one of the most powerful wavelet-based image compression method: the SPIHT (Set Partitioning in Hierarchical Trees)[11]. It is an image compression algorithm exploiting the inherent similarities across subbands in a wavelet decomposition

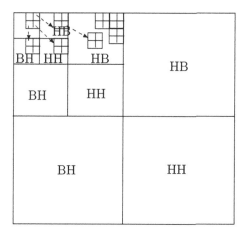

Fig. 26.3. Inter-subband dependencies used by the SPIHT algorithm

of an image. The SPIHT compression principle is based on the use of the zero-trees, in the wavelet subbands in order to reduce redundancies between them. Spatial orientation trees are created; they contain all the wavelet coefficients at the same spatial locations in the finer resolution subbands. Chap26figure3 shows an example of spatial orientation trees in a typical three level subband decomposition. The wavelet coefficients are encoded according to their nature: root of a possible zero-tree or insignificant set, insignificant pixel and significant pixel. The significance map is efficiently encoded by exploiting the inter-subband correlations and the bitplane approach is retained to encode the refinement bits. The SPIHT algorithm is mainly based on the management of three lists (List of Insignificant Sets, List of Insignificant Pixels and List of Significant Pixels). An iterative process successively scans and encodes the coefficients of each spatiotemporal tree[11].

26.5.2 Proposed image transmission system

The block diagram of the proposed image transmission system is given in chap26figure4. In this system, we use an image compression algorithm based on the DWT. he Lowest Frequency Subband LFS is coded separately from the Highest Frequency Subbands HFS. This allows unequal error protection to be easily applied. Also, if only a few levels decomposition are used, the decoded LFS would give a reasonable approximation of the entire image. The wavelet coefficients in the LFS are scalar quantized and then DPCM encoded. The latter is done by first finding a predicted value, for each coefficient, the prediction of a sample is merely the value of the previous sample. The predicted value is then subtracted from the coefficient to give residual coefficient, which is typically encoded. The HFS are encoded by the SPIHT algorithm.

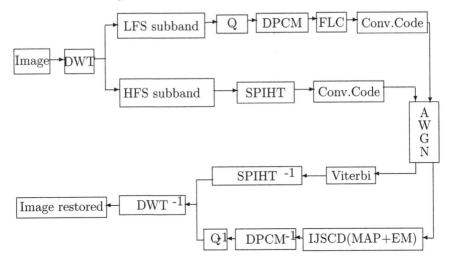

Fig. 26.4. Proposed image transmission system

The SPIHT coder provides good compression performance, but it is quite sensitive to bit errors. A convolutional code is then used for channel coding. The DPCM encoder leads to correlation among the transmitted indexes that can be considered as a first order Markov process. The idea is to apply the iterative joint source-channel decoding method, described in chap26section3, to data issued from the DPCM encoding of the LFS, in order to improve the image decoding. In the system that we propose, the wavelet coefficients of the LFS are scalar quantized and then DPCM encoded. Each obtained symbol is mapped into a binary code word. The resulting binary data are encoded using a recursive systematic convolutional code, and then transmitted over an AWGN channel. The iterative joint source-channel decoding method is applied at the receiver to decode the LFS data. The wavelet coefficients of the HFS subbands are coded by the SPIHT algorithm followed by a convolutional code. They are transmitted over the AWGN channel, then, they are decoded using the Viterbi algorithm followed by the SPIHT decoder. All subbands are regrouped. We finally apply a wavelet inverse transformation to restore the whole image.

26.5.3 Experimental results

In all our simulations, the Lena image of size $512 * 512$ pixels (8 bpp), is used as a test image. A three level wavelet decomposition is applied to the image, using the 9-7 filters. So the number of wavelet coefficients in the LFS is equal to 4096. These coefficients are quantized by a uniform scalar quantizer and then DPCM encoded. Each obtained symbol is represented by a 7-bit binary

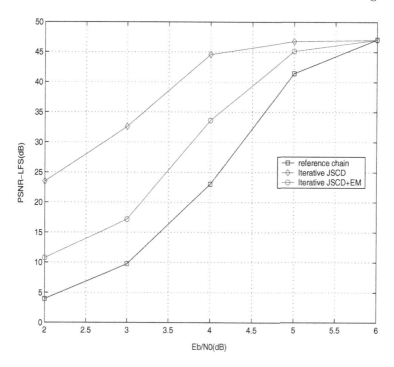

Fig. 26.5. PSNR of lowest frequency subband

code word. So we have a binary frame of 28672 bits representing the LFS data. This frame is shared into 32 packets (each packet has a length of 896 bits). These packets are then encoded by a recursive systematic convolutional code with generator polynomials $(37, 21)$ and rate $\frac{1}{2}$. They are transmitted over an AWGN channel. The IJSCD method, described in chap26section2 is applied at the receiver to decode the LFS data. The number of iterations is fixed to 3. The wavelet coefficients of the HFS subbands are coded by the SPIHT algorithm followed by a convolutional code with generator polynomials $(37, 21)$ and rate $\frac{1}{2}$. The rate at the output of the source coder is fixed to 1 bpp. The results are averaged over 500 channel realizations. In order to visualize the contribution of iterative joint decoding, we compared performance in terms of PSNR of the lowest frequency subband (PSNR-LFS), for the two systems without and with IJSCD decoding and source statistics estimation. Let's recall that our system applies IJSCD only to the LFS data, and that our reference system uses a separate source and channel decoding for both LFS and HFS data. The chap26figure5 represents the variation of the PSNR of the lowest frequency subband (PSNR-LFS) according to the signal to noise ratio $\frac{E_b}{N_0}$.

We can see that a significant gain in the PSNR of the LFS is obtained by using IJSCD. Indeed for a signal to noise ratio $\frac{E_b}{N_0} = 4dB$, we have a gain of about 8dB. The figure 6 represents the variation of the PSNR of the entire

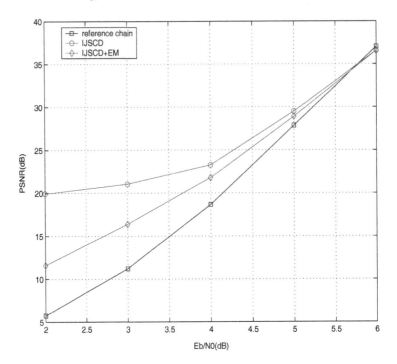

Fig. 26.6. PSNR of the entire image

image according to the signal to noise ratio $\frac{E_b}{N_0}$, for the image transmission systems without and with IJSCD and source statistics estimation. We can note that a significant gain in PSNR is achieved by iterative joint source channel decoding, comparing to classical decoding. This gain is about 1 dB for $\frac{E_b}{N_0} = 5dB$.

26.6 Conclusion

An efficient joint source channel decoding method, implemented by an iterative algorithm, and applied to an image transmission system is stated in this paper. The principle of this iterative algorithm is inspired from the turbo codes one; it uses the Baum Welch algorithm to estimate the source parameters at the receiver. A convolutional code is used for channel coding. Simulation show that, in case of a correlated gaussian source transmission, iterative joint source-channel decoding leads to a significant performance gain, in comparison with classical decoding. This iterative decoding scheme is applied to an image transmission system based on a wavelet transformation and a DPCM coding of the LFS and a SPIHT coding of the HFS. Channel coding is performed with a convolutional code. The simulation results indicate that the use of iterative joint decoding for the LFS data, can improve the error resilience of the image transmission system. The primary area of future research is improving the source compression, by using a variable length code instead of the fixed length one. We can also use turbo codes to improve the error protection.

References

1. T.Fine, Properties of an optimum digital system and applications, IEEE Trans.Inform. Theory, vol. IT-10 pp.287-296,1964.
2. J Garcias-Frias and John D.Villasenor, Joint Turbo Decoding and Estimation of Hidden Markov Sources, IEEE Selected Areas in Commun, vol. 19, pp. 1671-1679,2001.
3. C.Berrou, A.Glavieux, Near Optimum Error Correcting Coding and Decoding: Turbo-codes, IEEE Trans. Commun, vol.44, pp.1269-1271,1996.
4. N.Phamdo and N.Farvardin, Optimal detection of discrete Markov sources over discrete memoryless channels Applications to combined source-channel coding. IEEE Trans. Inform. Theory, vol. 40, pp.186-193,1994.
5. L.R.Bahl, J.Cocke, F.Jelinek, J.Raviv: Optimal Decoding of Linear Codes for Minimizing Symbol Error Rate, IEEE Trans. Inform. Theory, vol. 20,pp. 284-287,1974.
6. L.E.Baum and T.Petrie: Statistical inference for probabilistic functions on finite state Markov chains, Ann. Math. Stat, vol. 37, pp.1554-1563,1966.
7. L.E. Baum, T Petrie, G.Soules, and N.Weiss: A maximization technique occurring in the statistical analysis of probabilistic functions of Markov chains, Ann. Math. Statist., vol. 41, no. 1, pp. 164-171,1970.
8. J.Hagenauer, E.Offer, L.Papke: Iterative Decoding of binary Bloc and Convolutional Codes, IEEE Trans. Infor. Theory, vol. 42, pp. 429-445,1996.
9. J.Morlet: wavelett propagation and sampling theory, Geophysics,1982.
10. S.G.Mallat, A theory for multiresolution signal decomposition: The wavelet representation, IEEE Transactions on Pattern Analysis and Machine Intelligence, 1989.
11. A. Said and W.A.Pearlman, A new fast and efficient image codec based on set partitioning in hierarchical trees, IEEE Trans. Circuits Syst. Video Tech., vol. 6, pp. 243-250,1996.

AER Imaging

Mohamad Susli[1], Farid Boussaid[1] Chen Shoushun[2], and Amine Bermak[2]

[1] School of Electrical Electronic and Computer Engineering,
The University of Western Australia, Perth Australia,
[2] Department of Electrical and Electronic Engineering,
Hong Kong University of Science and Technology
Clear Water Bay, Kowloon, Hong Kong, SAR.

27.1 Introduction

Solid-state cameras have revolutionised the field of astronomy and are the enabling technology in the Hubble Space Telescope. Specially designed high-speed cameras have allowed scientists to visually observe time scales which were once an untouched domain. In the consumer market, they have made celluloid film-based cameras redundant and simultaneously created entirely new markets of their own. Indeed, solid-state imagers have become ubiquitous in all fields, driving the demand for low-power, high-performance imagers.

Despite its advances, a huge gap resides between what can be achieved in technology and what is achieved in biological sensors. The retina can adapt to almost any lighting condition and will consumes relatively little power. Engineers have turned their attention to biologically inspired camera architectures that promise to deliver high performance at an ultra-low power usage. This chapter will discuss one such avenue into "biomorphic" image sensors known as Address Event Representation (AER), starting with a brief background on image sensors in general.

27.2 Imager Basics

In digital cameras, images are acquired by a sequential readout from a photosensitive cell array [1]. The inferred brightness of each pixel will depend on the amount of light that falls on the photosensitive cell (otherwise known as a pixel) as well as on the duration of the integration period, the electronic analogue of the camera shutter speed.

A simplified physical circuit pixel circuit is shown in fig. 27.1, which includes a "reset" transistor that controls the integration period, and a junction capacitance which retains charge. As light with power P_{ph} strikes the photodiode, a current discharges the capacitor and reduces the voltage V_D shown in

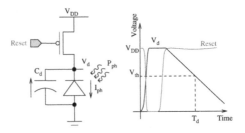

Fig. 27.1. The charge and discharge of a photodiode, operating as a pixel.

the diagram. Conventionally a circuit would wait until a time T_d and measure the final voltage on the capacitor. A lower V_D implies a brighter illumination incident on the pixel, since a larger current discharges the capacitor.

Conventionally, all pixels begin to integrate at the same time, and once halted an Analog to Digital Converter (ADC) must be used on each pixel sequentially. In the case of high resolution imaging systems, this conventional read out results in significantly reduced frame rate and leads to high power consumption since row and column pixel selection circuitry will need to be active for a longer period of time. Furthermore, faster or more numerous ADCs will need to be employed compounding the engineering problem and the power consumption issue.

Instead of integrating for a fixed time T_d, one can also wait for the voltage to reach V_{th} and measure the time taken. This has several advantages, most notably since it is simple to time an event using a digital circuit. If one was to construct a large array of such a pixel, given a random illumination profile, pixels will reach their threshold voltages at different times. Thus the problem of waiting for all pixels to complete their integration phase is circumvented. Furthermore, ADCs which consume a large amount of power in an imager are eliminated in favour of a timing circuit. The acquisition method described is referred to as Address-Event-Representation (AER) and is formally defined as a set of addresses (locations of pixels) each with a set of times representing events. This biologically inspired data representation is modeled after the transmission of neural information in biological systems [2][3]. An AER Stream, is a pulse data form signalling events as they occur. Using the information contained in an AER stream a simple signal processor can reconstruct the original image which created it.

Each time an event occurs (for instance when a predefined voltage is reached), a spike is generated by a pixel and a request for bus access is made to a peripheral arbiter. The latter takes the pixel address and places it on the bus. As a result, the asynchronous bus will carry a flow of pixel addresses. At the receiver end of the bus, address and time information are combined to retrieve the original data (e.g. pixel brightness value). In an AER-based imaging system, pixel read-out is initiated by the pixel itself. As a result, bus access is granted more frequently to active pixels (i.e., pixels that have generated events) than less active pixels, which will in turn consume much less communication bandwidth. The AER communication protocol makes efficient use of the available output bandwidth since read out can be achieved at any time upon request. In terms of power consumption, AER is also more efficient than the conventional fixed time-slot (synchronous) allocation of resources; this because not all pixels are likely to require computation/communication resources at the same time, hence there is no waste of resources. In the next section, we describe the basic building blocks of an AER based-imaging system.

27.3 AER-Based Imaging

27.3.1 Event Generator

In an AER-based imaging system a pixel comprises an event generator used to request access to the output bus each time a pixel has reached a predefined threshold voltage. The output of the event-generator pixel can be either a single pulse or a sequence of pulses [4]. In the latter case, the event-generator is referred to as Pulse Frequency Modulated (PFM) with the inter-spike interval a linear function of the pixel brightness value. In the case of a single output pulse, the event-generator is referred to as Pulse Width Modulated (PWM) because the duration of the pulse width is inversely proportional to the pixel brightness value. The PWM event-generator offers lower power consumption (a single transition) at the cost of a non-linear response [5]. Both PWM and PFM schemes encode illumination information in the time domain, providing noise immunity by quantisation and redundancy. In addition, representing intensity in the temporal domain, allows each pixel to have a large dynamic range (up to 200dB by modulating the reset voltage), since the integration time is not dictated by a global scanning clock. Moreover, time encoding ensures a relative insensitivity to the ongoing aggressive reduction in supply voltage that is expected to continue for the next generation of deep submicron silicon processes.

27.3.2 Arbitration

While offering many advantages, an AER based imager presents new problem to the designer as threshold events will be created in parallel. A feasible circuit would only be able to handle one such event at a time. There is always a likelyhood that two or more event in different locations will occur concurrently. This problem is referred to as a collision and an asynchronous circuit known as an "arbiter" is required to address this issue. The basic idea is to setup a queue where each pixel independently announces that it is ready to send its address (when the threshold is reached) and then awaits an acknowledgement from a control unit. When the acknowledgement is received, the pixel removes its request and resets itself in order to be ready for the next frame. The arbiter's job is to acknowledge only one of many requests, making it a critical element of the circuit. For example, In the situation where two requests come at the same time, only one of them will be chosen and acknowledged.

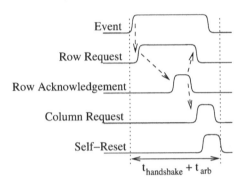

Fig. 27.2. The handshaking protocol of a single pixel with a horizontal time axis. The timing error is composed of an arbitration error and a handshaking error.

Since the pixels will be arranged in a grid, a topology must be utilised where pixels share rows and column "request" buses. In order for a pixel to be acknowledged, both its row and column requests must be acknowledged as illustrated by the simplified bus signals shown in fig. 27.2. Note that there are two types of timing errors. There is a signalling delay (handshaking error) labelled $t_{handshake}$ and a waiting error t_{arb} called the arbitration delay.

From the basic two-request line input arbiter shown in fig. 27.3.a), we can build up an arbiter for a bigger system by connecting the acknowledgement lines into another arbiter's request line. The generation of these signals is split up into arbitration, propagation and acknowledgement units. The arbitration unit selects one of two request lines and stores its decision in an internal state. The propagation unit will then send a signal to the next stage of arbitration, which when finished will send an acknowledge signal back to the arbiter as shown in 27.3.b. The arbiter will then send an acknowledge signal to the pixel chosen by its internal state.

This way the outputs of two arbiters become the input of one arbiter, forming a four-request line arbitration block. Since each arbitration block will have its own delay time, it is trivial to show that this delay will be logarithmic[3] with respect to the number of request lines. This delay is the arbitration delay t_{arb} mentioned previously and it is critical that it should be minimised in order to increase the image quality in the camera.

---

[3] This is of course if we assume all delays are equal, normally two temporally close requests will generate a larger delay.

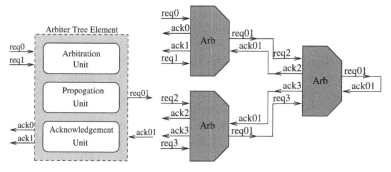

Fig. 27.3. a) A 2-input building block of a hierarchical arbitration tree. b) A 4-input arbitration tree.

27.4 Event driven Simulation

An AER simulator was developed to evaluate the performance of possible AER image coding implementation schemes. The simulator was developed in C++. It has very efficient runtime performance when compared to similar simulators utilising Verilog, requiring less than a seconds[4] to simulate the behaviour of a 128x128 pixel array. The block diagram of the simulator is shown in fig. 27.4. An image is initially loaded by the "loading stage" and converted from intensity information into timing by a reciprocal relationship. Since digital representation of intensity have no units, a constant of proportionality τ is used. The second stage involves simulating the behaviour of a physical AER imager, with arbitration and row/column delays. Finally a computational stage calculates the overall reduction in quality created by the arbitration process.

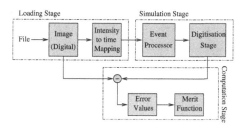

Fig. 27.4. Block diagram of simulator.

The AER simulator system uses an event-wheel based on the heap data structure. This event-wheel stores the upcoming events such as a pin going high, an arbiter receiving an edge or a photodiode discharging down to its threshold voltage. The heap data structure insures that the events are

---

[4] As tested on a x86 based 1.73GHz processor.

processed sequentially in time. However, the event-wheel has additional functionality which processes some events with higher priority. In this design, processed events will generate events in the future. These are pushed on to the event-wheel, which will continue to process events until it is empty, in which case the simulation ends. The initial state of the camera is assumed to be a reset state, where no pixels are discharging and all buffers are cleared. An image is then loaded, whose digital values are translated into discharge to threshold times (also known as Time to First Spike or TFS). A global integration signal is then simulated, where all pixels create an event in the future at the time when they are expected to reach their threshold voltage. The state of the pixels are thus in the "integrating state" and will eventually be processed one by one into their next state.

Since the simulator is event-based, even though these processes are simulated sequentially, it is effectively concurrent. The current time is never incremented manually; instead it is refreshed with the latest event's value. By operating in this fashion, delays will propagate themselves and thus simplify the operation of the simulator greatly. The buffers hold the locations of the pixels which have fired and are awaiting arbitration. This is to simulate the memory realised by the physical buffers. This buffer structure is routed into the neighboring arbiters using a look up table, based on row or column location. This speeds up simulation immensely, however it needs to be initialised once before simulation can commence.

The arbiters accurately simulate the logical behaviour of the real cell. There are two events which effect row and column arbiters, "arb_edge" and "arb_update". The former is an event which is called whenever an arbiter's neighboring cell changes its output. Since it takes some time window in order to decide on the arbitrated signal (if any), some propagation time is taken before the new even "arb_update" is generated. This will then also have a propagation delay of its own, depending on whether the priority or non-priority line has been selected. This models the real life behaviour, since the pull strength of the non-priority line is weaker, it will take longer to create a transition. Finally, arbiters toggle priority by simply swapping a logic bit upon selection of a line. When the simulation ends, each pixel will store a value which contains the total time until it receives a column acknowledgement. This is taken to be equivalent to the time which it would be placed on the AER stream. Thus by creating a new heap and pushing on these locations and times, an AER stream is formed by this simulator. Finally, to reconstruct the image either a time to intensity mapping or a simulated timing circuit can be used. Using this, a comparison can be made between the original image, and the errors generated by arbitration and AER architecture.

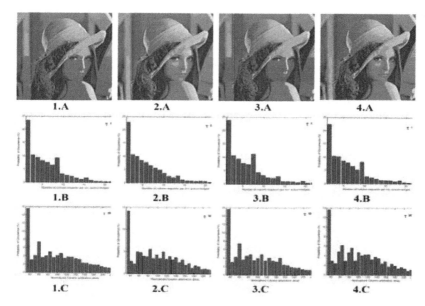

Fig. 27.5. Fixed arbitration (1.A-C and 2.A-C) and Fair arbitration (3.A-C and 4.A-C): Delay and number of requests as a function of the intensity-to-time conversion factor . The original image Lena is 256×256.

27.5 Results

27.5.1 Latency

The AER simulator emulates the finite time required for arbitration of an image. It thus introduces additional timing errors as illustrated in fig 27.5. In AER, brighter pixels are favoured because their integration threshold is reached faster than darker pixels. As a result, brighter pixels will request the output bus more often than darker ones. This results in an unfair allocation of the bandwidth. Two different arbitration schemes were examined, with images 1 and 2 utilising a fixed arbiter, and images 3 and 4 utilising a fair arbiter. In the fixed arbiter, priority is given always to the same input lines. In contrast, a fair arbiter toggles the priority between input lines. It should be noted that fair arbitration gives a subtle enhancement of the output image. In addition to different arbitration schemes, different values of τ were used in order to evaluate the degradation of image quality, representing the conversion factor used by the AER simulator to map intensity into the time domain (First block in 27.4).

This effect is quite pronounced, with the upper half of the image gaining priority over the lower half. An AER imager typically processes its columns after a row has been chosen to have a greater priority over the others which are waiting. Thus the time it takes to process any row is compounded over all waiting rows. In order to evaluate this timing delay, a histogram (images 1B to 4B) was constructed consisting of the number of pixels per row which make a request once their row has been acknowledged. The results (which were cropped to eliminate negligible bins) show that fair arbitration tends to reduce the amount of requests on average, but only a small amount for this image. Finally, the delay of each row was recorded, and plotted in a histogram representation, for each respective image. As intuitively expected, there is a direct correlation between the number of pixel requests and the delay, on average.

27.5.2 Event Generator

In fig. 27.6, we see the difference between PWM and PFM encoding, with different values of τ. Both of these simulations are run using fair arbitration. Figures 1-3A show the PWM encoding scheme at values of the intensity-to-time factor τ of 1.1, 0.7 and 0.3 respectively. We see that for a value of 1.1, arbitration causes only a few minor losses in contrast. Lowering τ to 0.7, the error begins to dominate in one half of the image and at 0.3 the loss in contrast is apparent everywhere in the image. We can see the error due to PFM encoding in figures 1-3B, for the same values as before, however with a capture time of 0.1. This value represents the amount of time the simulation is run upon which all events are blocked and the occurrences are counted immediately.

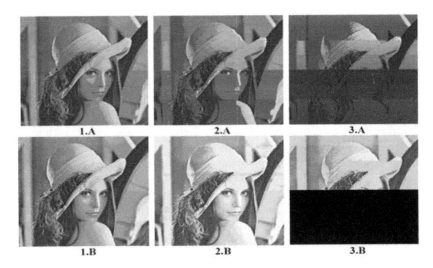

Fig. 27.6. PFM and PWM encoding as a function of the intensity-to-time conversion factor τ. The left image represents the highest value of τ with proceeding images representing a lower value.

For a value τ of 1.1, the image quality is greater than that of the PWM encoding scheme when judged by a human observer. However, the image has lost a great deal of contrast. At a value of 0.7, the image loses more contrast, however the quality can be judged to remain high than that of the TFS image. Finally at a value of 0.3, the imager begins to suffer from arbiter starvation, where a single row arbiter is receiving requests faster than it is dropping them, leaving one half of the image without acknowledgement.

27.6 Conclusion

In this chapter, we have explored the potential benefits of AER image coding schemes for high speed low power image capture and transmission. An in-house AER simulator was developed to examine potential arbitration schemes such as fixed and fair arbitration protocols and evaluate their performance evaluated in terms of event generator element, latency and output bit stream compression.

References

1. Fossum, E.: CMOS Image Sensors: Electronic camera-on-a-chip. IEEE Tran. Electron Devices **44** 1997 1689–1698
2. Sivilotti, M.: Wiring considerations in analog VLSI systems with applications to field programmable networks. Ph.D. dissertation, California Institute of Technology, Pasadena, 1991
3. Mahowald, M. A.: VLSI analogs of neuronal visual processing: a synthesis of form and function. Ph.D. dissertation, California Institute of Technology, Pasadena, 1992
4. Culurciello E., Etienne-Cummings R. and Boahen K. A.: A biomorphic digital image sensor. IEEE Solid-State Circuits, **38** 2003 281–294
5. Kitchen A. , Bermak A. , and Bouzerdoum A.: A Digital Pixel Sensor Array With Programmable Dynamic Range. IEEE Trans. Electron Devices **52** Issue 12 2005 2591–2601
6. Boahen K. A.: Point-to-point connectivity between neuromorphic chips using address events. IEEE Trans. on Circuits and Systems II, **47** Issue 5 2000 416–434
7. Culurciello E., Etienne-Cummings R., and Boahen K.: Arbitrated address-event representation digital image sensor. Electronics Letters, **37** Issue 24 2001 1443–1445
8. Rullen V. and Thorpe S. J.: Rate Coding Versus Temporal Order Coding: What the Retinal Ganglion Cells Tell the Visual Cortex. Neural Computation, **13** 2001 1255–1283

Large deviation spectrum estimation in two dimensions

Mohamed Abadi and Enguerran Grandchamp

GRIMAAG UAG, Campus de Fouillole
French West Indies University
97157 Pointe--Pitre Cedex Guadeloupe France
{mabadi,egrandch}@univ-ag.fr

Summary. This paper deals with image processing. This study takes place in a segmentation process based on texture analysis. We use the multifractal approach to characterize the textures. More precisely we study a particular multifractal spectrum called the large deviation spectrum. We consider two statistical methods to numerically compute this spectrum. The resulting spectrum, computed by both methods over an image, is a one dimension spectrum. In the scope of this article, we extend these methods in order to obtain a two dimensions spectrum which could be assimilated to an image. This 2D spectrum allows a local characterization of the image singularities while a 1D spectrum is a global characterization. Moreover, the computation of the spectrum requires the use of a measure. We introduce here a pre processing based on the gradient to improve the measure. We show results on both synthetic and real world images. Finally, we remark that the resulting 2D spectrum is close to the resulting image of an edge detection process while edge detection using one dimension spectrum requires post processing methods. This statement will be used for future works.

Key words: multifractal analysis, multifractal spectrum, numerical computing spectrum, Hlder exponent, Choquet capacity

28.1 Introduction

Texture analysis techniques have been intensively studied over the last decades, among and the image processing community. Within these techniques, multifractal analysis was introduced by Parisi and Frisch [3] to study the singularities of 1d-signals and has yielded some interesting results. Nevertheless, as many tools of multifractal analysis have been developed initially for 1d-signals, there is no direct way to use them on images without loosing the intrinsic 2d-relation between two neighbour pixels. For example, [16] used the large deviation spectrum to detect edges in images. However, in this study, the computation of the large deviation spectrum considers the image as a 1d-signal.

This article deals with the generalisation of large deviation spectrums to the case of 2d-signals. In order to do so, we will reconsider many approaches from the 1d-case. All of these approaches deal with a so-called multifractal spectrum which is roughly a tool used to quantify the number of points having the same Hölder exponent (singularity). As the estimation of this number of points is particularly difficult when dealing with discrete data, many numerical approaches can be found in the literature.

The original study [3] was based on the study of the power law behaviour in structure functions [6], [7]. As the computation used the Legendre transform, the estimated multifractal spectrum was called "Legendre Spectrum". However, as shown by Muzy and al. [8], Arneodo and al. [9], the structure function method has many drawbacks. Particularly, it does not allow to access to the whole spectrum. They both present a new method to apply a multifractal analysis based on a wavelet transform modulus maxima [10],[11],[12] still conducting to a Legendre spectrum estimation.

Other authors suggest applying the multifractal analysis on a measure defined over the signal itself. Turiel and al [13],[14] compute fractal sets and are particularly interested on the MSM (Most Singular Manifold) set. MSM allows to characterize a signal from a geometrical and statistical point of view applying the gradient operator over the initial signal and then using a wavelet transform in order to determine the fractal sets.

Lévy-Véhel and al. [15] use the Choquet capacity firstly to define measures, secondly to determine the Hölder exponents and then to compute the multifractal spectrum. In this way, they introduce the kernel method and the histogram method to estimate, in a one dimension context, a multifractal spectrum called the "large deviation spectrum" [1]. This spectrum allows to characterize the singularities in a statistical way.

This last approach, as previously said, was applied successfully in [16] to an application of edge detection and is the one we would like to generalise.

The article is built as follow. After having presented some mathematical pre-requisite and the way to compute the singularity exponents and 1d large deviation spectrum (section 28.2) we will focus on the 2d case (section 28.3) in which the resulting spectrum is an image. As the spectrum computation depends on the definition of a measure, we will test two of them. The first uses the Choquet capacity as in [15], [18] and we will introduce a second measure based on the combination of the gradient and Choquet capacity. A comparison between the results obtained with each measure will be made in section 28.4. Section 28.5 is dedicated to conclude the article.

28.2 Multifractal formalism

We present in this section the formalism used to compute the multifractal large deviation spectrum. We use the following steps:

1. Image normalization,
2. Multifractal measure defined by the Choquet Capacity [15],
3. Hölder exponents computation,
4. Spectrum computation.

28.2.1 Singularities computation

Let μ be a measure defined over a set $E \in [0, 1[\times [0, 1[$, $P(E)$ is a partition sequence of E and ν_n is an increasing sequence of positive integer.
    In this case, the partitions are defined as follow:

$$E_{i,j,n} = \left\{ \left[\frac{i}{\nu_n}, \frac{i+1}{\nu_n} \right[\times \left[\frac{j}{\nu_n}, \frac{j+1}{\nu_n} \right[\right\}$$

For image analysis applications, we choose that the set $E_{i,j,n}$ is a window of size n centred on the point of coordinates (i, j), i.e. $|E_{i,j,n}| = n$. This window is slide over the whole image by moving the center to its neighbours. In other words, the centre of the new set $E_{i',j',n}$ will have the coordinates $(i', j') = (i + 1, j + 1)$ if the movement is over the image diagonal, $(i', j') = (i, j + 1)$ for a horizontal one and $(i', j') = (i + 1, j)$ for a vertical one. $(i', j') = (i + 1, j)$ for a vertical one.
    Then for each image point (i, j) singularities exponents are given by the Hölder exponents.

$$\alpha(x, y) = \lim_{r \to \infty} \frac{\log[\mu(B_r(x, y))]}{\log(r)}$$

Where $B_r(x, y)$ is a window of size $r = 2m + 1$ with $m = 0, 1, \cdots, \lfloor \frac{n}{2} \rfloor$ and $(x, y) = (1, \cdots, r)^2$.
    $|E_{i,j,n}|$ is the size of the partition of E and μ the measure defined by the Choquet capacity on each window. figure 28.1. shows a representation of an image and three windows, respectively of size $r = \{1, 3, 5\}$.
    In practice $\alpha(x, y)$ is determinate by the slope of the linear regression of the following log curve: $\log[\mu(B_r(x, y))]$ versus $\log(r)$. The Figure 28.2. shows the projection of the measure, built in figure 28.3. with a sum operator capacity, over the logarithmic scale and also the singularity computation using the slope of the linear regression $(\alpha(i, j) = 2.288)$. This allows to characterize the behaviour of the measure μ at the neighbourhood of (x, y).
    For image processing applications, the multifractal analysis is based on the estimation of the multifractal spectrum determined by the Hausdorff dimension [19], the Legendre spectrum [15] or the large deviation spectrum [1]. In the scope of this article we study the last spectrum.

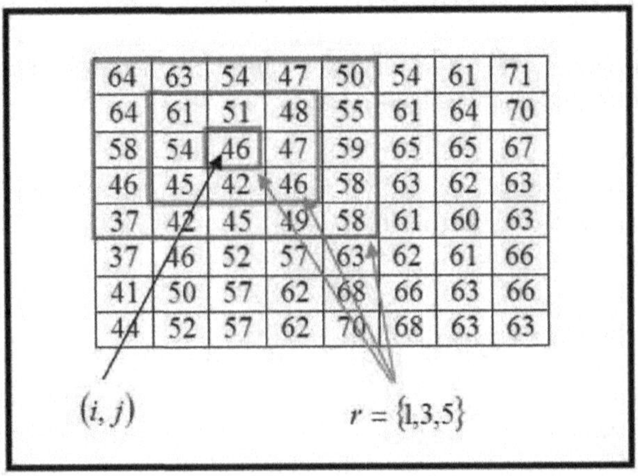

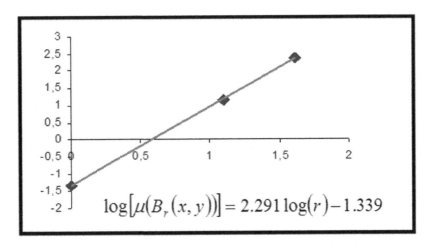

64	63	54	47	50	54	61	71
64	61	51	48	55	61	64	70
58	54	46	47	59	65	65	67
46	45	42	46	58	63	62	63
37	42	45	49	58	61	60	63
37	46	52	57	63	62	61	66
41	50	57	62	68	66	63	66
44	52	57	62	70	68	63	63

(i, j) $r = \{1,3,5\}$

Fig. 28.1. Matrix representing the image and three windows respectively of size $r=1,3,5$

$$\log\left[\mu\left(B_r(x, y)\right)\right] = 2.291 \log(r) - 1.339$$

Fig. 28.2. Linear regression on a logarithmic scale

The main idea is to use a sequence of Choquet capacities which allows the extraction of local and global information from the image in order to study the singularity behaviour.

28.2.2 Choquet capacity measure

In this section, μ is a measure defined by the Choquet capacity. In the literature we found many capacities [16], [17] with a general definition having the following shape:

0	0	0	0	0	0	0	0
0	0	0	0	0	0	0	0
0	0	2.291	2.284	1.867	1.846	0	0
0	0	2.613	2.368	1.934	1.919	0	0
0	0	2.311	2.232	1.983	1.995	0	0
0	0	1.971	1.972	1.899	1.996	0	0
0	0	0	0	0	0	0	0
0	0	0	0	0	0	0	0

Fig. 28.3. Hölder coefficients after image normalization with $r = \{1, 3, 5\}$ and $n = |E_{i,j,n}| = 5$

$$\mu(x, y) = O(i, j)_{\in B_r(x,y)} \, g(i, j)$$

With O an operator dealing with the intensity of a pixel $g(i, j)$. As examples, we can cite: the sum operator $O = \sum$, which is not a real informative measure of the image since it computes the sum of the intensities within a window, the maximum and minimum operator respectively $O = \max$ and $O = \min$, which have a low sensibility to the singularity amplitude. Other operators have been introduced like self-similar or iso operator, more details are given respectively in [18] and [15].

The main drawback of these operators is their lack of sensibility to the amplitude or to the spatial distribution of the singularities.

In this article, our gait takes as a starting point the work carried out by Turiel and al. [13] to determine the fractals sets. We combine one of the previous operators with the gradient ∇ computed on each pixel, defined over two axes, and the norm. Thus we obtain three measures which are sensible simultaneously to amplitude and spatial distribution of the singularities. These measures have the following expression

$$\mu_x(x, y) = O\nabla_x g(x, y)$$

$$\mu_y(x, y) = O\nabla_y g(x, y)$$

$$\mu_{xy}(x, y) = \sqrt{[\mu_x(x, y)]^2 + [\mu_y(x, y)]^2}$$

Using these measures we can compute the singularity coefficients along the two axes and also that the norm. In this paper, we use, in particular, the gradient norm because it allows a correct representation and describe the brusque variations of images intensity:

$$\alpha_x(x, y) = \lim_{r \to \infty} \frac{\log[\mu_x(B_r(x, y))]}{\log(r)}$$

$$\alpha_y\left(x,y\right) = \lim_{r\to\infty} \frac{\log\left[\mu_y\left(B_r\left(x,y\right)\right)\right]}{\log\left(r\right)}$$

$$\alpha_{xy}\left(x,y\right) = \lim_{r\to\infty} \frac{\log\left[\mu_{xy}\left(B_r\left(x,y\right)\right)\right]}{\log\left(r\right)}$$

After the computation of the Hölder exponents, we can focus on the multi-fractal spectrum estimation. In the following of the article, we will study the definition and the method to compute the large deviation spectrum.

28.3 Numerical estimation of the large deviation spectrum

Let us introduce in this section a two dimension adaptation of the two methods defined by Lévy Véhel and al. [1]. This adaptation allows estimating the large deviation spectrum from a measure construct by a combination between the previous operators and the gradient computed on both axes and previously describing.

This is a way to characterize the singularities and to study their behaviour in a statistical point of view. In the two dimension case, we define the large deviation spectrum as follow:

$$f_g\left[\alpha\left(i,j\right)\right] = \lim_{r\to\infty} \frac{\log\left[N_r\left(\alpha\left(i,j\right)\right)\right]}{\log\left(r\right)} \quad (M1)$$

$$f_g^\varepsilon\left[\alpha\left(i,j\right)\right] = \lim_{\varepsilon\to 0}\lim_{r\to\infty} \frac{\log\left[N_r^\varepsilon\left(\alpha\left(i,j\right)\right)\right]}{\log\left(r\right)} \quad (M2)$$

where $N_r\left[\alpha\left(i,j\right)\right] = \#\left\{\alpha\left(x,y\right)\ /\ \alpha\left(i,j\right) = \alpha\left(B_r\left(x,y\right)\right)\right\}$ for the first method and $N_r^\varepsilon\left[\alpha\left(i,j\right)\right] = \#\left\{\alpha\left(x,y\right)\ /\ \alpha\left(B_r\left(x,y\right)\right) \in \left[\alpha\left(i,j\right) - \varepsilon, \alpha\left(i,j\right) + \varepsilon\right[\right\}$ for the second one, which is a variant. $\alpha\left(i,j\right)$ is the singularity in the centre of the window B_r of size r,

$\alpha\left(x,y\right)$ is the singularity within B_r at the spatial coordinates $\left(x,y\right)$.

The first estimation using $(M1)$ allows to compute the number $N_r\left[\alpha\left(i,j\right)\right]$ of singularities $\alpha\left(i,j\right)$ equals to $\alpha\left(B_r\left(x,y\right)\right)$. For the second estimation $(M2)$, $N_r^\varepsilon\left[\alpha\left(i,j\right)\right]$ represent the number of $\alpha\left(x,y\right)$ that belong to the interval $\left[\alpha\left(i,j\right) - \varepsilon, \alpha\left(i,j\right) + \varepsilon\right[$.

For image processing purpose, both methods are summarized with the following algorithm:

for each pixel (i,j),

for $m = 0$ to $m = |E_{i,j,n}|$

$$r = 2m + 1$$

compute $N_r\left[\alpha\left(i,j\right)\right]$ (resp. $N_r^\varepsilon\left[\alpha\left(i,j\right)\right]$)

There is three particular values of m

$m = 0 \Rightarrow r = 1 \Leftrightarrow B_{r=1} = 1$ pixel$\Leftrightarrow (x, y) = (i, j)$ (minimal window size)

$m \neq 0$ and $m \neq |E_{i,j,n}| \Leftrightarrow B_r$ is a window of size $r \times r \Leftrightarrow (x, y) \in \{1, 2, \cdots, r\}^2$

$m = |E_{i,j,n}| \Leftrightarrow B_r$ is a window of size $|E_{i,j,n}| \times |E_{i,j,n}|$ where $(x, y) \in \{1, 2, \cdots, |E_{i,j,n}|\}^2$ (maximum window size)

The spectrum will be estimated by the slope of the linear regression $\log[N_r(\alpha(i,j))]$ versus $\log(r)$. figure 28.4. illustrates the Hölder exponents and three windows used to compute the number of singularities $N_r(\alpha(i,j))$ centred on (i, j). Figure 28.5. shows the projection over the logarithmic scale and the linear regression for both methods and also the computation of the large deviation spectrums f_g et f_g^ε .

Then figure 28.6. shows the large deviation spectrum matrices for both methods.

0	0		0	0	0	0	0
0	0	0	0	0	0	0	0
0	0	2.291	2.284	1.867	1.846	0	0
0	0	2.613	2.368	1.934	1.919	0	0
0	0	2.311	2.232	1.983	1.995	0	0
0	0	1.971	1.972	1.899	1.996	0	0
0	0	0	0	0	0	0	0
0	0	0	0	0	0	0	0

Fig. 28.4. Hölder coefficients and window of size $r = 1, 3, 5$

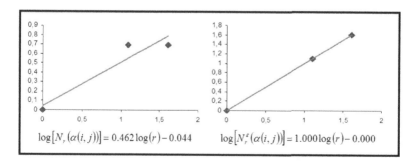

$$\log[N_r(\alpha(i,j))] = 0.462 \log(r) - 0.044$$

$$\log[N_r^\varepsilon(\alpha(i,j))] = 1.000 \log(r) - 0.000$$

Fig. 28.5. large deviation spectrum estimation with two methods $\varepsilon = 0.3$

0	0	0	0	0	0	0	0
0	0	0	0	0	0	0	0
0	0	0.462	0.462	0.000	0.000	0	0
0	0	0.000	0.000	1.041	0.883	0	0
0	0	0.000	0.000	0.000	0.833	0	0
0	0	0.674	1.000	0.000	0.000	0	0
0	0	0	0	0	0	0	0
0	0	0	0	0	0	0	0

0	0	0	0	0	0	0	0
0	0	0	0	0	0	0	0
0	0	1.000	1.095	1.136	1.136	0	0
0	0	1.462	1.195	1.534	1.407	0	0
0	0	1.000	1.534	1.534	1.512	0	0
0	0	1.095	1.381	1.557	1.348	0	0
0	0	0	0	0	0	0	0
0	0	0	0	0	0	0	0

Fig. 28.6. large deviation spectrum estimation with two method matrices with $n = 5$

28.4 Results and experiments

In this section, we apply the two previous methods for the large deviation spectrum estimation over a synthesis image (Figure 28.7) and also over an image extracted from the FracLab software (Figure 28.8). Then we compare the measure that we introduce with the other measures (Figure 28.9, 28.10).

Figure 28.7 shows that it is interesting to introduce the gradient before applying an operator. In fact the three lines are underlined after the computation of the singularity exponents.

Figure 28.8. shows the singularity results with and without gradient. Singularities seem richer when using the gradient.

The more interesting comparison is shown in figure 28.9 and 28.10. The first notable result is the display of a two dimensional spectrum. The figures show a better spectrum obtained with the gradient operator. Concerning the two methods used to compute the spectrum, we notice a better result with the second one due to ε.

28.5 Conclusion and future works

This study, deals with large deviation spectrum estimation in two dimensions. The first main conclusion is that the measure based on the gradient that we introduce is an efficient way to improve intensity variations detection. The second main conclusion is that the large deviation spectrum estimate on each pixel according to its neighbours gives a local and a global characterization of the information.

Large deviation spectrum is widely used for segmentation in the following way: computation of the singularity, computation of one dimension spectrum, segmentation of the image by integrating spectrum and singularity. Our approach allows to directly obtain a two dimension spectrum which is closed to

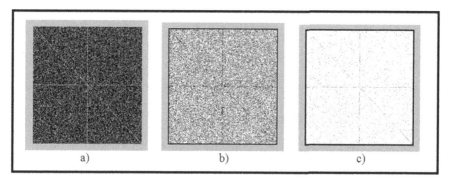

Fig. 28.7. a) Image representing three lines (horizontal, vertical and diagonal) with a gaussian noize ($\sigma = 0.6$). b) Hölder coefficients with the iso capacity. c) Hölder coefficient computed with the gradient operator followed by the $O = iso$ capacity here ($n = 5$).

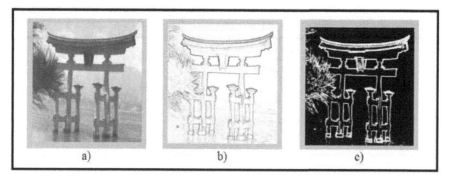

Fig. 28.8. Original image extracted from the FracLab software [2]. b) Singularity exponents computed with the min capacity ($n = 3$). c) Singularity exponents computed with the gradient operator followed by the $O = sum$ capacity, with $n = 3$.

segmentation. It will be interesting to compare the two segmentation results. In the same way, the introduction of the gradient before integrating a one dimension spectrum will be compared with two dimension spectrum.

In addition by using the second method based on the ε−value can be improve by defining a criterion of optimization which allows giving the ε_{opt} optimal value is under development.

This spectrum has been estimated using two methods based on measures built using Coquet capacity. It will be interesting for classification and segmentation purposes to combine these different spectrums (one spectrum per measure) in order to qualitatively show the interest of this study.

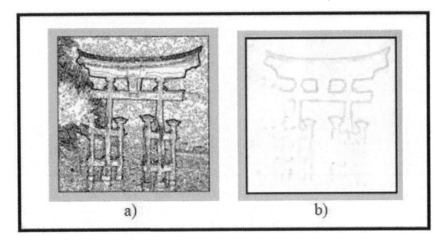

Fig. 28.9. a) Large deviation spectrum estimated using the first and the second approach ($\varepsilon = 0.2$) with $n = 7$ from the singularity exponents of Figure 28.8 b).

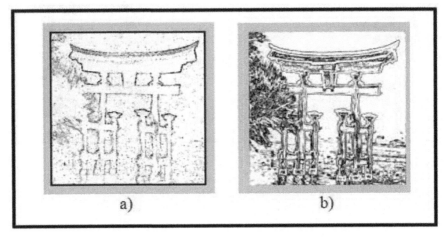

Fig. 28.10. a) Large deviation spectrum estimated using the first and the second approach ($\varepsilon = 0.2$) with $n = 7$ from the singularity exponents of Figure 28.8 c).

Acknowledgments

The authors would like to thank the European institutions for the financing of the CESAR (Arborescent species classification) project and Guadeloupe, Martinique and Guyana regions within the "INTERREG IIIb Caribbean Space" European program

References

1. J. Lévy Véhel, Numerical Computation of Large Deviation Multifractal Spectrum, In CFIC96, Rome, 1996.
2. http://www.irccyn.ec-nantes.fr/hebergement/FracLab/
3. G. Parisi, U. Frisch, Turbulence and Predictability in Geophysical Fluid Dynamics and Climate Dynamics, Proc. of Int. School, 1985.
4. A. S. Monin, A. M. Yaglom, Statistical Fluid Mechanics, MIT Press, Cambridge, MA, vol. 2, 1975.
5. A. S. Monin, A. M. Yaglom, Statistical Fluid Mechanics, MIT Press, Cambridge, MA, vol. 2, 1975.
6. A. S. Monin, A. M. Yaglom, Statistical Fluid Mechanics, MIT Press, Cambridge, MA, vol. 2, 1975.
7. U. Frisch, Turbulence, Cambridge Univ. Press, Cambridge, 1995.
8. J. F. Muzy, E. Bacry, A. Arneodo, Phys. Rev. E 47, 875, 1993.
9. A. Arneodo, E. Bacry, J. F. Muzy, Physica A 213, 232, 1995.
10. A. Grossmann, J. Morlet, S.I.A.M. J. Math. Anal 15, 723, 1984.
11. A. Grossmann, J. Morlet, Mathematics and Physics, Lectures on Recent Results, L. Streit World Scientific, Singapour, 1985.
12. M. B. Ruskai, G. Beylkin, R. Coifman, I. Daubechies, S. Mallat, Y. Meyer, L. Raphael, Wavelets and Their Applications, Boston, 1992.
13. A. Turiel, N. Parga, The multi-fractal structure of contrast changes in natural images: from sharp edges to textures, Neural Computation 12, 763-793, 2000.
14. A. Turiel, Singularity extraction in multifractals : applications in image processing,
    Submitted to SIAM Journal on Applied Mathematics.
15. J. Lévy Véhel, R. Vojak, Multifractal analysis of Choquet capacities, Advances in applied mathematics, 1998.
16. J. Lévy Véhel, P. Mignot, Multifractal segmentation of images, Fractals, 371-377, 1994.
17. J.-P. Berroir, J. Lévy Véhel, Multifractal tools for image processing, In Proc. Scandinavian Conference on Image Analysis, vol. 1, 209-216, 1993.
18. H. Shekarforoush, R. Chellappa, A multi-fractal formalism for stabilization, object detection and tracking in FLIR sequences.
19. J. Lévy Véhel, C. Canus, Hausdorff dimension estimation and application to multifractal spectrum computation, Technical report. INRIA, 1996.

Index

Continued from page ii